To Alfred Tenny,
with warm regards,
Roger Jones Sipi
Glück auf!

Alfred —
Enjoy the book!
Sue Wulcoviz

The Chemical Industry and Globalization

ACS SYMPOSIUM SERIES **942**

The Chemical Industry and Globalization

Roger F. Jones, Editor
Franklin International, LLC

Sponsored by the
ACS Division of Business Development and Management

American Chemical Society, Washington, DC

Library of Congress Cataloging-in-Publication Data

American Chemical Society. Meeting (229th : 2005 : San Diego, Calif.)
  The chemical industry and globalization / Roger F. Jones, editor ; sponsored by the ACS Division of Business Development and Management

   p. cm.—(ACS symposium series ; 942)

  Revised papers of a symposium sponsored by the Division of Business Development and Management at the 229th national meeting of the American Chemical Society held in San Diego, California, March 13–17, 2005.

  Includes bibliographical references and index.

  ISBN 13: 978–0–8412–3977–7 (alk. paper)

  ISBN 10: 0–8412–3977–0 (alk. paper)

  1. Chemical industry. 2. Globalization.

  I. Jones, Roger F. II. Title. III. Series.

HD9650.5.A46 2006
338.4′766—dc22
                                                             2006042790

The paper used in this publication meets the minimum requirements of American National Standard for Information Sciences—Permanence of Paper for Printed Library Materials, ANSI Z39.48–1984.

Copyright © 2006 American Chemical Society

Distributed by Oxford University Press

All Rights Reserved. Reprographic copying beyond that permitted by Sections 107 or 108 of the U.S. Copyright Act is allowed for internal use only, provided that a per-chapter fee of $33.00 plus $0.75 per page is paid to the Copyright Clearance Center, Inc., 222 Rosewood Drive, Danvers, MA 01923, USA. Republication or reproduction for sale of pages in this book is permitted only under license from ACS. Direct these and other permission requests to ACS Copyright Office, Publications Division, 1155 16th Street, N.W., Washington, DC 20036.

The citation of trade names and/or names of manufacturers in this publication is not to be construed as an endorsement or as approval by ACS of the commercial products or services referenced herein; nor should the mere reference herein to any drawing, specification, chemical process, or other data be regarded as a license or as a conveyance of any right or permission to the holder, reader, or any other person or corporation, to manufacture, reproduce, use, or sell any patented invention or copyrighted work that may in any way be related thereto. Registered names, trademarks, etc., used in this publication, even without specific indication thereof, are not to be considered unprotected by law.

PRINTED IN THE UNITED STATES OF AMERICA

Foreword

The ACS Symposium Series was first published in 1974 to provide a mechanism for publishing symposia quickly in book form. The purpose of the series is to publish timely, comprehensive books developed from ACS sponsored symposia based on current scientific research. Occasionally, books are developed from symposia sponsored by other organizations when the topic is of keen interest to the chemistry audience.

Before agreeing to publish a book, the proposed table of contents is reviewed for appropriate and comprehensive coverage and for interest to the audience. Some papers may be excluded to better focus the book; others may be added to provide comprehensiveness. When appropriate, overview or introductory chapters are added. Drafts of chapters are peer-reviewed prior to final acceptance or rejection, and manuscripts are prepared in camera-ready format.

As a rule, only original research papers and original review papers are included in the volumes. Verbatim reproductions of previously published papers are not accepted.

ACS Books Department

About the Editor

Editor Roger F. Jones is president of Franklin International LLC, a management consulting firm offering services to the chemical and plastics industries, and non-executive chairman of the board of directors of PlastiComp, LLC, a new technology development and licensing firm. In his 50-year career in the plastics industry, he has been president of Franklin Polymers, LNP Engineering Plastics, and Inolex Chemical Company, managing director of BASF Engineering Plastics, a group executive with Beatrice Chemical, and held management/professional positions in R&D, marketing, and manufacturing with DuPont, Avisun, and ARCO Chemical (both now parts of BP Chemical).

He is a graduate of Haverford College, holding a BS with Honors in Chemistry and Honorable Mention in English Literature. He is a Fellow of both the Society of Plastics Engineers and the American Institute of Chemists. He is a 50-Year Member of the American Chemical Society and a member of Sigma Xi. He retired from the U.S. Navy Reserve with the rank of captain, after 34 years of service.

Contents

Preface xi

1. **The Chemical Industry in the 21st Century** 1
   Roger F. Jones
      Is the U.S. Losing Its Manufacturing Base? 2
      U.S. Manufacturing and China 6
      The U.S. Chemical Industry 11
      Employment: Overall and Chemistry Professionals 16
      The Escalation of Competition 22
      Industry and the Environment 23
      Globalization and Regional Markets 25
      Influence of the Financial Community 27
      Duplicative or Differing Visions? 31
      Is U.S. Chemical R&D in Decline? 34
      What Might Be Done 38
      Conclusions 41
      References 43

2. **Research and Development in the Pharmaceutical Industry and Investment in Innovation** 47
   Susan Wollowitz and Faiz Kermani
      Introduction 48
      Innovation Is the Life Blood of the Pharmaceutical Industry 48
         Public Health Benefits from Innovation 52
         Regions Want Pharmaceutical R&D 53
         Venture Capitalists and Start-Up Enterprises 54
      Constraints on Innovation: Cost Containment—
         Public Health and Public Benefit Collide 55
      Intellectual Property Rights: When Innovation Cannot
         Be Turned into Profit 56
      Effect of Globalization on R&D Activity 56
      Trans-Oceanic Dynamics 58
      U.S. R&D Dominance 60
      State Investment Is Critical 65
      California—Home of Biotech 67
      Massachusetts—Another Innovation Driven Hotbed 68
      States That Purposefully Create Innovation Centers 70
      U.S. Private Investor Contribution to Stimulating R&D 72

European Complexity 73
Mounting Cost Pressures 76
Boosting Pharmaceutical Innovation in Europe 76
The Environment for R&D in the UK 80
The Environment for R&D in France 83
The Environment for R&D in Germany 86
The Environment for R&D in Spain 88
Japan 90
Emerging Asia 92
India Invests for the Future 93
Singapore 95
China 96
Stem Cell Research—An Example of Global Competition 99
Alternate Models of Investment 101
Conclusions 103
References 104

3. The China Challenge 111
Timothy C. Weckesser

The Big Picture: Economic Trends and Implications 112
The World's Workshop 113
Automobiles 115
Telecommunications 117
Software 120
Power and Pollution 124
The Environmental Challenge 131
Rise of the Private Sector 134
The Entrepreneurial Challenge 138
The Rise of Chinese Multinationals 141
Key Market Entry Issues 144
Intellectual Property Protection 144
Guanxi 148
Homework 150
Differentiation and Pricing 151
Short- and Long-Term Strategy 152
Summary 154
Postscript 155
References 159

Indexes

Author Index 163

Subject Index 165

Preface

Although the science of chemistry has been globalized for much more than 50 years, the chemical industry has only felt the full effects of globalization in the past five to ten years. This has been largely due to the rise of China as a major industrial power. The subject of globalization is not only very broad but also constantly changing so that portions of this book will be unavoidably dated by the time of publication. The authors have tried to depict the major aspects of globalization and how they have impacted various regions of the world and industry sectors. Information that has been published in the popular press and even in industry publications during the past several years has been fragmentary and conflicting, occasionally inaccurate or even misleading. The authors have attempted to organize and analyze publicly available data to present a more comprehensive, factual, and useful description of the impact of globalization on our industry and our profession.

Three principal areas of interest are addressed in this book: the chemical industry (excluding pharmaceuticals), the pharmaceutical industry, and the business situation inside China. Past, present, and future trends in the employment of chemical professionals are examined, as this affects most American Chemical Society (ACS) members directly. We hope the reader will find this review useful to understanding what is happening both to the chemical industry and to the people who work in it.

The Business Development and Management Division and the Committee for Economic and Professional Affairs cosponsored the symposium at which these papers were presented, at the 229th National Meeting of the ACS in San Diego, California, March 13–17, 2005. The symposium sponsors did not require papers to be submitted, and the authors have expanded significantly upon what they presented, adding more recent information.

The editor and authors thank the symposium sponsors for providing a forum to explore this important subject and the ACS Books Department for its desire to publish our work. Finally, we owe special thanks to the manuscript reviewers, Ernest C. Coleman, Thomas C. Gates, Peter L. Lantos, and Tracy E. Rusch whose comments were invaluable to the authors for making the book more accurate, readable, and meaningful.

Roger F. Jones
Franklin International LLC
4 Kenny Circle
Broomall, PA 19008
FranklinIntl@aol.com (email)

The Chemical Industry and Globalization

Chapter 1

The Chemical Industry in the 21st Century

Roger F. Jones

**Franklin International LLC, 4 Kenny Circle, Broomall, PA 19008
(email: FranklinIntl@aol.com)**

Over the past decade the maturing of some sectors of the chemical industry, the acceleration of manufacturing productivity growth, globalization, and the growing power of Wall Street over business strategies, has been changing the nature of the industry significantly. While the bubble economy of the 1990s tended to mask many of the effects, they have now become painfully clear in the face of unprecedented run-ups in the cost of natural gas and crude oil derived feedstocks. We face a very different business environment in the 21st century than we did just five years ago. However, it is important to note that all is not what it seems. A number of the perceived problems are overstated, obscuring those worth our concern. Some of the problems are regional while others are global. Some are lasting and others transitory. This paper attempts to establish the facts that offer more insightful knowledge of the current status of the chemical industry than one would learn from uninformed and alarmist accounts in the popular media. It also attempts to analyze which circumstances are important and those that are transitory, what has brought us to this point, the outlook for the future, and what corporate management can do to succeed under these conditions.

© 2006 American Chemical Society

Is the U.S. Losing Its Manufacturing Base?

There is said to be an ancient curse, "May you live in interesting times." Nothing could be more "interesting" than what is happening to the chemical industry as we enter the third millennium. All manufacturing is changing in the U.S., and the chemical industry is caught up in these broad changes as well as in ways that are specific to it; the "rules" have changed. To make things harder for all involved, the rate of change itself appears to be accelerating, as foretold in Alvin Toffler's famous 1970 book, *Future Shock*. Using the conventional descriptive labels applied to chemical industry, some say that "commodity" chemicals have finally become mature in the classical economic sense of not growing faster than the GDP rate, but this is only true in Western countries and Japan. (*1*) However, specialty chemicals, including polymers and plastics, exhibit significantly higher growth than the GDP rate, and the top of the ladder is most certainly occupied by pharmaceuticals. Nevertheless, even these growth areas are almost all marked by increasingly intense competition, creating management emphasis on restructuring and reengineering, all of which usually results in a reduction of jobs. Unfortunately, there are a number of misperceptions about what is actually taking place, particularly those stemming from the more sensationalist stories in popular news sources. It is important to separate the myth from the reality before one can ever begin to understand the problems that globalization presents and deal effectively with them.

Over recent decades, the single most important cause for the disappearance of manufacturing jobs is not – as popularly supposed – the "export" or "outsourcing" or "offshoring" of jobs to Mexico, China, or other overseas locations, but rather the result of growing productivity improvement. Advances in productivity are essential to increasing industry competitiveness – and the faster productivity gains come, the faster they also shrink the overall number of manufacturing jobs, not just locally but globally as well. This is not a new process by any stretch of the imagination; it has been going on for many years. For example, both U.S. and China manufacturing employment declined during the period 1995-2002, 11% in the U.S. (this was also the global average), but a much higher 15% in China, partly due to the rationalization of some of the more inefficient state-owned firms. (*2*) In the course of the 2000-

2003 U.S. recession, many people confused the highly publicized loss of U.S. manufacturing jobs at the time with the equally well-publicized rapid growth of the Chinese manufacturing sector and erroneously concluded that U.S. manufacturing *per se* was being transferred to China on a massive scale. This focus on China's rapid development overlooked the fact that most of the U.S. job losses by that time had already taken place (at the beginning of the U.S. recession when's businesses reacted to the sharp downturn in demand by laying off unneeded workers). Although China was enjoying a major domestic economic expansion in most of this same period, the additional jobs created in 2000-2002 only restored China's manufacturing employment level to barely that of 1998. (*3*)

Another factor that has mislead people as to what is taking place in U.S. manufacturing, is that the service sector of the economy has been growing much faster than manufacturing for a number of years. The relative percentage of the GDP representing manufacturing has thus declined, from 18.5% in 1992, to 14.1% in 2002, (*4*) leading to the fallacy that manufacturing is declining on an absolute basis, not just a relative one. For quite some time now, the press has regularly played up Department of Commerce reports of a negative balance of trade in manufacturing when it occurs, but has typically failed to point out that those same Department news releases also usually report an accompanying *positive* balance of trade in services.

It has been disappointing to see even scientists willing to accept isolated anecdotal instances as proof of a hypothesis, particularly those published in the popular press – a secondary source at best. Easily accessible evidence clearly shows that U.S. manufacturing is both vital and growing, quite likely *because* it has been shedding unneeded jobs and becoming ever more efficient. During 1977-1997, the most recent pre-recession period available as of July 2005, U.S. Census Bureau (*5*) data plainly show that overall manufacturing has continued to *expand*, not shrink, within the U.S. (see Figure 1):

- Number of firms *up* 3 %
- Productivity *up* 147 % (1987-2004)
- Employment *down* 9 %
- Sales *up* 282 %
- New capital investment *up* 292 %

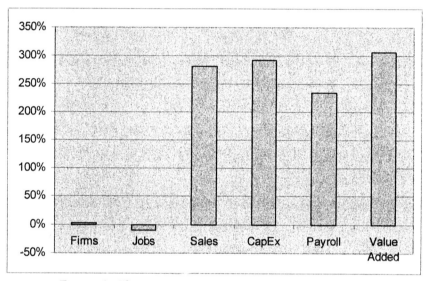

Figure 1. Changes in U.S. Manufacturing, 1977-1997
Source: U.S. Census Bureau

Yes, there were fewer jobs, and industry consolidation has offset most, but certainly not all, new company formations. The data unmistakably show that there were *more* companies, making and selling *more* goods, and making *more* capital investments by the end of this period, with sales and investment having grown at average annual rates *in excess of 14%*. Lest the reader suspect that the best paying jobs were being moved overseas, the data show manufacturing payrolls over this period increased 235%, or 11.8% annually – since actual employment was declining, the average manufacturing worker's income was gaining substantially, indicating that if any jobs were outsourced, they were the lower paying ones. Likewise, it appears that manufacturing has been shifting from mature to newer products, because the value added by manufacturing rose by an even greater rate, 307%, 15.4%/year.

Since the average GDP rate during this same period was 3.2%, per the U.S. Bureau of Economic Analysis (BEA) (*6*), this can hardly be called the picture of an entire sector of the economy being outsourced. In addition to these facts, it is worth noting that U.S.-based manufacturing output has not only remained the largest in the world, but is two and half times the size of the second largest, Japan. While Germany is considered to be the third largest manufacturing economy in the world, its economy is essentially stagnant; China's manufacturing sector is growing rapidly and may well overtake Germany in the near future. (*7*)

This is not to say that U.S. manufacturing is not having problems or could not do better. Some sectors have certainly been hurting badly while others have done quite well – that is the trouble with averages, they don't reveal how much variance exists. Nevertheless, the problems that affect everyone need to be looked at closely. The National Association of Manufacturers (NAM) has shown that U.S. federal government policies are a substantial hindrance for U.S. manufacturers competing against offshore companies (*8*). It may surprise the reader to learn that, compared to the U.S.'s nine largest trading partners as of December 2003, U.S. corporate income tax rates and costs of environmental and safety regulatory compliance were the highest of any. The U.S. also has higher government-mandated employee benefit costs than six of these largest trading partners, with only France, Germany, and South Korea being more than those in the U.S. It should come as no surprise that tort liability costs in the U.S. (2.23% of the GDP) dwarf those anywhere else in the world. Asbestos liability litigation has been a particularly pernicious example, with over 8400 firms being named defendants by plaintiffs, 75% of whom have no detectable symptoms of asbestosis and even though asbestos ceased being used for industrial purposes more than thirty years ago. The sum of these government-imposed additional cost burdens works out to about $5 per employee per hour worked, nearly equal to the *entire* Chinese manufacturing cost burden.

The National Association of Manufacturers (NAM) has other issues with the way the Federal Government treats manufacturing, too, such as the lack of an Assistant Secretary for Manufacturing in the Commerce Department. Because of NAM's initiative, such a position was created and filled in early 2005; whether or not this will result in any positive outcome remains to be seen. One hopes that the new bureaucrat will be working toward less government involvement rather than more. However, this would be atypical – the temptation to pick industry winners and losers has been the historical outcome of such activities and is therefore more likely. One suspects after some experience with "our man in the government," NAM will find that inviting the government to be involved in making industrial policy was a dubious idea, and one they will regret demanding.

U.S. Manufacturing and China

China's low labor costs have been widely blamed as the main reason for the seeming outsourcing of U.S. manufacturing. Obviously, this is a matter of "comparative advantage," at which the Chinese presently excel vs. the U.S. However, many in U.S. industry believe that China's manufacturing costs have been artificially enhanced by the currency valuation policies of its government. For several years, NAM and other groups have urged the U.S. Government to persuade or force China to revalue its currency upwards as much as 40%, to reduce this presumed unfair trade advantage. (Please bear with the following – the July 2005 change in the valuation of the yuan is dealt with at the end of this section.) From an economist's point of view, this demand has little, if any, recognizable merit. For one thing, even though the World Trade Organization (WTO) encourages members to float their currencies, it does not consider pegging the value of one currency to another to be a violation of trade laws, so there really are no legal underpinnings to a "demand" for a change. China observed first hand that the Asian financial meltdown in 1998 was initiated by Thailand and other Asian countries revaluing their currencies at the urging of western nations (including the U.S.). As a result, China has sought to value stability above all in its financial markets. China's weak banking system, which holds an enormous amount of uncollectible loans (estimated by a number of economists to be as much as half the total), would be seriously impacted by any major currency revaluation. Consequently, China has preferred to deal with the question of export competitiveness indirectly, such as through export taxes, rather than change the set value of the yuan against the dollar.

No country has ever derived any lasting benefit from the manipulation of its currency valuation – it is simply too easy for other nations to do the same, and trade wars generally end badly for all concerned. Despite the lessons of history, the idea still remains popular among economists and politicians who are sure that "this time it will be different," although it never is. The same demands were placed on Japan over 15 years ago, to revalue its currency upwards (the yen's value has long been "managed" by the Japanese government's intervention in currency markets). The Japanese reluctantly went along with the idea and many economists point to this event as the leading reason for the decade-long recession and deflation that followed.

One idea that has been floated in some circles, including Congress, is to place a punitive duty (27.5%) on Chinese imports, to compensate for the alleged unfair currency valuation advantage. This scheme should be considered dead on arrival. Not only would this surely be ruled illegal by the WTO, but other nations would then be permitted by the WTO to levy retaliatory tariffs on U.S.-made products. U.S. exporters are already facing heavy retaliatory tariffs as a result of the WTO declaring the Byrd Amendment to be illegal (it encourages U.S. companies to complain of "dumping" under rules that make it a difficult charge to disprove, and if the claim is upheld, these same companies are allowed to keep *all* of the proceeds of punitive tariffs that are imposed). These are by no means the only instances where the U.S. has not lived up toWTO rules, and there is a serious risk that continually doing so will lead to an escalating trade war – one thinks back to the global one created by the 1930 Smoot-Hawley tariffs and which played a major role in turning a recession into the economic disaster known as the Great Depression.

If the Chinese were to revalue the yuan upwards by a large amount, would this actually have a net benefit for U.S. manufacturers? Imported components that many firms buy or make in China would then become more costly. Would this result in the repatriation of any outsourced production? It would seem more likely that the now more costly Chinese goods would be simply replaced by less expensive products from other developing countries, e.g., India or Brazil. China has also become a major importer of U.S. goods: exports to China have grown from $15B in 2001 to $30B to 2003 (9). While U.S. goods would become less expensive as the result of revaluation, China's economy would be likely to cool down with the result that it would import less. Alan Greenspan, while still Federal Reserve Chairman, said that even a 10% upward revaluation of the yuan would not be likely to have any significant positive impact for U.S. manufacturers.

Another unanswered question in the event that Chinese exports to the U.S. were to contract significantly is, would the Chinese continue to passively invest their trade surplus dollars in U.S. Treasuries or would they begin to look for something offering greater financial returns? The U.S. government's spending deficit is financed through issuing these very same U.S. Treasury securities – if the demand for them were to contract, U.S. interest rates would very likely rise, which would adversely affect the U.S. economy, particularly the housing market. For the Chinese, the alternative to buying U.S. Treasuries would be to buy U.S. stocks or acquire companies.

Several Chinese companies have already made efforts to acquire U.S. firms, e.g., Lenovo has acquired IBM's personal computer business unit, Heier has tried (unsuccessfully) to buy Maytag appliances, and Chinese National Offshore Oil Company (CNOOC) tried to acquire Unocal. All of these companies have varying degrees of Chinese government ownership, which has raised questions about unfair financing advantages (e.g., non-recourse and below-market interest rate government loans) over U.S. companies bidding on these same acquisitions. In the case of CNOOC, the reaction from both U.S. industry and Congress was so strongly negative that CNOOC withdrew its offer. Since CNOOC is 70.8% owned by the Chinese government, national security issues concerning the availability of oil from Unocal in the event of a conflict cannot be easily dismissed. Perhaps a bid by a truly privately owned Chinese company would not arouse such a strong response.

Much of the controversy over exchange rates and the increased level of imports from China seems a bit overblown when one realizes that the U.S. imports from Pacific Rim countries as a group reached a peak of 39% of total imports in 1993 and have been declining ever since (down to 34% in 2003) (*10*); see Figure 2. During this time, China's share of U.S. imports has indeed doubled from 6 to 12%, but this has been more than offset by a drop in imports from other Asian countries. There is scarcely anything sinister or even unusual about such a relatively normal ebb and flow of trade over a ten year period – the fuss being made at present is strongly evocative of the hue and cry raised in the late 1980s, warning how "Japan Incorporated" was taking over the U.S. through rising exports and asset purchases, a scare that swept the country, even leading to movies and books on the subject. The Japanese economic invasion was never effectively consummated, certainly not on any scale that might have been genuinely troubling. This may also be the same story for China. Nevertheless, there are very significant differences between Japan and China: the former has a free enterprise economic system under a democracy that has been functioning successfully for over 50 years and has been genuinely friendly to the U.S., while the latter has a mixed state-private enterprise system under a politically repressive, non-democratic Communist government – one of the few left in the world – and one that views the U.S. as its primary opponent in the world. The directions given by the Chinese government to the companies it controls are not transparent and therefore lead to reasonable concern about what their ultimate objectives might be.

All of this said, the picture changed on July 21, 2005, when the Chinese government surprised the world when it announced that it was changing the yuan from its ten-year long pegged rate against the dollar to a "controlled float." The Chinese accomplished a brilliant diplomatic coup by appearing to accommodate the pressure from the U.S. and EU governments, while actually doing very little. The float has been instituted with an initial 2.1% revaluation, hardly more than a token concession. While the conditions of the float specify using an unnamed basket of currencies for valuation but only permitting a movement of 0.3% per week, currency traders have said that the Chinese central bank has been intervening heavily in the time since the evaluation was announced, restraining further appreciation of the yuan to not more than 0.2% through September 2005. This suggests that the yuan exchange

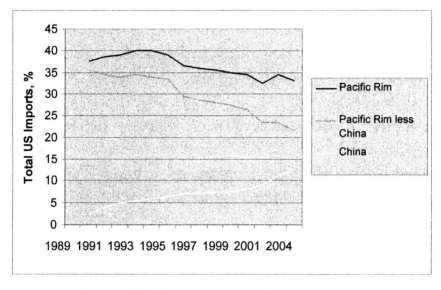

Figure 2. U.S. Imports, Pacific Rim, and China
Source: U.S. Bureau of Economic Analysis 2005

rate will remain very much just whatever the Chinese government wants it to be, rather than be determined by currency markets. (*11*) Following revaluation, U.S. economics professors and business reporters were finally being quoted in the press that a more expensive yuan may well create as many or more problems than it solves, e.g., mortgage rate increases and a slowing, if not the end, of the U.S. housing boom, etc. (*12, 13*)

The U.S. Chemical Industry

The U.S. chemical industry has been doing better than the rest of U.S. manufacturing, but faces some extraordinary difficulties in the years to come. Unfortunately, U.S. Census data (*14*) for the various sectors of manufacturing, including the chemical industry, are only currently available for the years 1997-2002 (as of July 2005), so we cannot make an exact direct comparison of how the chemical industry has fared vs. overall manufacturing through 2004. However, the data are useful for getting a picture of the proportion the chemical industry comprises of total manufacturing as well as how the industry fared during the recession. Looking at 1997 data as representative of a non-recession year, one finds that chemical companies represented 2.9% of the total number of manufacturing firms, but while chemical industry jobs were 5.3% of total manufacturing employment, they represented 7.0% of total manufacturing payrolls. This suggests that the chemical industry tends to be more concentrated in fewer firms than the average, but as we shall see in a minute, this is somewhat misleading. In any event, it is clear that jobs in the chemical industry are higher paying than the average for manufacturing overall. Chemical industry sales made up 10.8% of total manufacturing sales and chemical industry capital expenditures represented 14.3% of the total. So we can say that, at least in 1997, the chemical industry was not only a high value component of U.S. manufacturing, but also a significant buyer of capital goods.

In 2002, the recession was in full swing and much hand-wringing was observed in the press. Despite the decline in domestic demand, crude oil and natural gas prices surged to "record" levels (although inflation-adjusted comparisons for oil prices show that these were far from record prices – exceeding the record for crude oil would require inflation-adjusted prices over \$93/bbl., and the prices being reported were really "futures," not actual trades). The drop in demand, coupled with the rapid escalation in raw material costs caused high stress in the chemical industry. Industry executives labeled this situation as the most challenging in fifty years. However, looking at the 2002 data cited earlier, we find something unexpected: chemical companies represented 3.8% of the total manufacturing firms, a 31% increase – evidently fewer chemical firms went out of business or merged than other types of manufacturers did during the recession. Likewise, chemical manufacturing sales rose to 11.8% of the total (very likely due to the unexpectedly successful pass-through of sharply increased raw material costs). Plus chemical industry employment rose to 5.8% of total manufacturing employment, and chemical industry payrolls grew to 7.7% of the total. Only capital expenditures contracted more than overall manufacturing, dropping to 11.9% of the total. *In every category but one, the chemical industry outperformed overall manufacturing during the recession;* see Figure 3. Thus, it is demonstrably unjustified to speak of the U.S. chemical industry as greatly weakened or even disappearing from the U.S. manufacturing sector. To the contrary, the industry has successfully met the test and moved forward. The future in the U.S. looks good, too: Klaus-Peter Löbbe, president of BASF North America, recently said that his company views North America as the largest market for chemicals in the world over the next ten years (granted that his public relations department has also announced several North American plant closures and attendant employment cutbacks to improve profitability). [*36*]

Despite this sterling performance, the U.S. chemical industry has many serious challenges ahead of it. First and foremost is not simply the substantial increase in domestic raw material costs in the past few years, but a looming significant and long-term shift in regional cost advantages. Relatively abundant and inexpensive natural gas has provided a competitive advantage to those U.S. chemical companies that use C_2 and C_3 olefin-based chemical processes for many decades. This advantage has almost completely disappeared in the dramatic run-up of natural gas prices during the past several years. The reason for the escalation in prices is the direct outcome of U.S. government policies that have encouraged the use of natural gas as fuel for power generation while restricting the development of new sources and pipelines for distribution within the U.S. While the availability of biomass sources for raw materials may ease some actual shortages, their costs are sufficiently high that users cannot compete readily in world markets against petrochemical feedstock users. The same holds true for coal

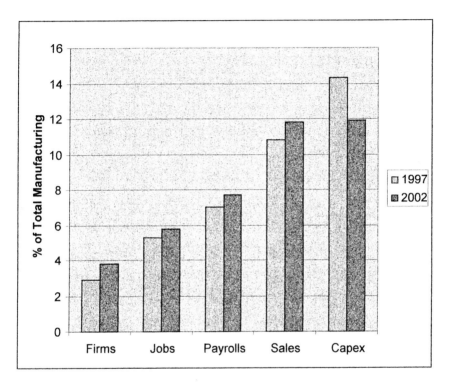

*Figure 3. U.S. Chemical Industry vs. Overall U.S. Manufacturing
Growth Year (1997) vs. Recession Year (2002)
Soruce: U.S. Census Bureau*

gasification – costs are too high to be competitive in world markets. Both sources are seeking public subsidies from Congress, but this shows that they are not yet ready to compete on an even playing field and, if they are subsidized, would distort economic signals from the marketplace as to whether they actually offer a realistic alternative.

Chemical uses of natural gas hardly appear to be on the radar screen of anyone outside of the chemical industry. Congress reluctantly confronted the supply problem but the legislation that was enacted in mid-2005 will do little to reduce restrictions on exploration and development of offshore gas fields, only calling for the Secretary of the Interior to compile a catalogue of the existing fields. Exploration and development of new coastal areas and Alaska's Arctic National Wildlife Refuge (ANWR) were not included in the final legislation. It is most regrettable that many members of both political parties in Congress complain that the U.S. is overly dependent on foreign oil but then simultaneously adamantly refuse to support the exploration and development of new and promising domestic sources.

Importation of liquefied natural gas (LNG) might be helpful but cannot be considered a meaningful short-term alternative; the amount that can be supplied is minimal with only two terminals in service. The outlook for more terminals is at best uncertain due to the time required to obtain the necessary permits for building terminals, nominally four to seven years, including construction time. Even this lengthy period can be and is frequently extended by opposition from communities and environmental groups, citing potential explosion and fire hazards associated with handling such products. The new energy legislation, mentioned earlier, seeks to improve this situation by giving exclusive authority to the U.S. government over offshore LNG terminals, taking state and local authorities out of the approval process. Some forty LNG terminal applications will be affected.

Even if exploration and drilling restrictions had been lifted by this legislation, significant increase of domestic oil and gas supplies would be unlikely before 2009, again owing to permitting and construction lead times. Well before this time, massive competitive olefin manufacturing capacity, now under construction in the Middle East, will be brought onstream (it was originally scheduled for 2006 but has slipped to 2007-2008 or even later due to construction delays). *(15,16)* These plants will be based on huge local natural gas fields – Saudi Arabia's and Iran's reserves each are only exceeded by those of Russia; in other words, they are numbers two and three in the world. In the past, these Middle Eastern countries have flared their natural gas as waste, so their cost basis is considered merely that of transportation and storage. These new plants, totaling several billions of tons of olefin capacity, will result in a major, long-term, and very likely permanent shift of economic advantage to this region of the world for commodity olefin-based products. The Middle East producers have identified Asia as their primary market, but they will obviously displace suppliers now based in North America and Europe.

U.S. polypropylene (PP) and polyethylene (PE) producers have been fortunate that demand in China has been so high in the past several years that they have been able to export there at U.S. domestic prices. However, this situation is already beginning to fade as the Chinese government's efforts to cool down the country's superheated demand have taken effect. China also has substantial domestic polyolefin capacity now coming onstream, as well as more in planning or under construction. In view of the impending loss of cost advantage, most of U.S. producers have been deferring or outright canceling significant chunks of their planned domestic polyolefin plant expansions. There are a few exceptions to this generalization: Total Petrochemicals, for example, has announced plans to build a new, world-scale, 300,000 tons per annum (K TPA) polypropylene line in the U.S. by 2008-2009, on top of a 100 K TPA expansion just completed in 2004 at its Texas plant. While these numbers may seem impressive in isolation – and they are big for Total, the world's fourth largest PP producer – they will nevertheless add only a few percentage points to overall North American polypropylene capacity. (*17*)

Hurricane Katrina and, to a lesser extent, Hurricane Rita, have significantly damaged the Alabama-Mississippi-Louisiana-Texas Gulf Coast areas at the time this book was written. These are possibly the most severe back-to-back natural disasters to befall the U.S. in the past 100 years. While damage to this major area of gas and oil extraction, refining, and chemical manufacturing sites does not appear to have been as extensive as it might have been, the immediate outcome has been a further dramatic escalation of natural gas prices. This has occurred because gas supplies, already tight, have been yet further constrained by extended shutdowns of Gulf Coast oil and gas fields, as well as refineries, caused by these two major storms. With winter approaching, it appears it will be quite some time before inventories can be rebuilt to pre-storm levels. Corporations have taken a hard line with customers, insisting that they cannot absorb these raw material cost increases and must pass them through immediately. The net effect will be to make U.S. producers even more vulnerable to future Middle East imports. Chemical industry executives have warned Congress that, without meaningful and immediate easing of restrictions on increasing natural gas supplies, much of their business will be lost, and it will be extremely difficult to get it back. The industry is looking hard at alternative raw material bases that are less subject to the price volatility and supply shortfalls that have come to characterize oil and gas in the past few years (although industry leaders claim that as much as 25% of the observed price volatility in oil and gas comes from the activities of speculative hedge funds).

European olefin producers have both benefited and been hurt by the strong Euro-weak dollar situation of the last several years. Unlike the U.S., the primary feedstock for olefins in Europe is oil, which is cracked to produce olefins. Oil is priced in dollars, so that European producers have seen somewhat less run-up in cost than in the U.S. and other dollar zone economies, such as China. While this is a benefit for business done within the Euro zone, the European Union's (EC) largest countries, such as Germany, France, and Italy, have had minimal economic growth in the past several years because domestic demand has been very weak. Countries such as Germany, which has traditionally depended on exports for almost a third of its GDP, have been severely hampered by the strong Euro in non-Euro markets, a second significant factor in low GDP growth rates. While the newer EC states, such as Poland, offer potentially higher growth markets within the Euro zone, they are still relatively small and it will be some time before they grow to a size that will take enough export sales to make a difference in the "old Europe" GDP rates. This has meant that many European chemical producers are consequently investing outside Europe, particularly in China and India in order to have a significant future for their industry, both short and long term.

Employment: Overall and Chemistry Professionals

A closer look at U.S. government employment statistics vs. mass media reports also shows surprising differences, akin to the ones we have observed on manufacturing. As mentioned earlier, one problem with the popular press is that layoffs are always deemed newsworthy but hirings far less so. The press has seemed particularly uninterested in reporting the reduction in unemployment and upward swing in job creation following the recovery from the recession, 2003 to the present, and has commonly *mis*reported the findings of the U.S. Bureau of Labor Analysis (BLA). BLA "conducts two monthly surveys that measure employment levels and trends: the Current Population Survey (CPS), also known as the household survey, and the Current Employment Statistics (CES) survey, also known as the payroll or establishment survey." (*18*)

As of July 2005, the CPS household survey was based on a sample of 60,000 households, seeking to find out how many in each household surveyed were employed. The small sample size vs. the millions of households nationally means that only a minimum month-to-month difference of ± 436,000 can be considered statistically significant. The data are also adjusted annually for population growth. The CES payroll survey is based on a monthly sample of 400,000 business and government establishments. This larger sample in a smaller population results in a much smaller month-to-month statistically meaningful minimum difference, ± 108,000. The primary difference between the two employment data sources is that the CPS measures small business and self-employment, particularly at the startup stage, while the CES measures employment at larger, established businesses. BLS states that the two surveys measure different aspects of employment in the economy and therefore must be considered together to understand employment trends.

Nevertheless, it is notable that CPS data are almost completely ignored by the press when reporting employment trends, who also persist in publishing changes in the monthly employment numbers that are not statistically meaningful per the above definitions (e.g., changes of 30,000), but without noting this important fact. Why is using both surveys so important? The reason is clear when one sees that the trough-to-peak (November 2001-December 2004) CES total employment gain was 1,395,000, while the CPS total employment gain was an astonishing *3,752,000*! Not reporting these latter gains hugely distorts the employment picture during this period.

One may also discern other trends from looking closely at these unreported data that cannot be found in the "popular" CES survey. The CPS survey is where small company startups first show up – typically entrepreneurs are most active within a year or so after an economic recovery begins. Many economists are aware that small company data are notoriously underreported by government agencies; these firms are hard to find because they are often unincorporated sole proprietorships. While it is a strictly anecdotal example, the author has been the CEO at four small companies for a total of 19 years, dating as far back as 1970, and none of his firms were ever surveyed. According to the most recent BLS data (November 2002), small companies (under 250 employees) employ 55% of all workers in the chemical industry (*19*); see Figure 4. Thus we have surprisingly little reliable information to tell us whether *total* chemical industry employment is actually declining or not. Considering that most large companies have been consolidating and restructuring, it would seem to be a reasonable assumption that any real employment growth must be taking place almost entirely in small and new companies at present, and hardly at all in most of the large, established ones.

Reliable data about the distribution of chemical professionals by company size do not appear to exist. Perhaps the best indication we have of overall chemistry professional employment trends is the American Chemical Society's (ACS) annual employment survey. This survey has shown an increase in unemployment among chemists in the past several years, but participation in this survey is voluntary and is solicited only from its approximately 155,000 members. While the ACS survey questionnaire mailing includes the estimated 9300 chemical engineers who are ACS members, this is much smaller than the 40,000 members of the American Institute of Chemical Engineers (although there may be some overlap), and therefore may not be representative of chemical engineers overall. As a group, the survey indicates that chemical engineers, compared to chemists, appear to be more concentrated in industry, earn higher salaries, are more likely to be employed full-time and less likely to be unemployed. The ACS survey therefore may be neither statistically reliable nor necessarily representative of the chemistry profession as a whole.

"According to the 2004 annual ACS survey of employment status of members in the domestic workforce (2004), 3.6% of them were unemployed but seeking employment. This was a record high in the 30-plus-year history of the surveys, if only by 0.1% over the previous high in 2003. The percentage with full-time jobs reached an all-time low of 90.9%. Of the remainder, 3.6% were employed part time and 1.9% were on postdocs or fellowships." (*20*) This same survey showed that "about

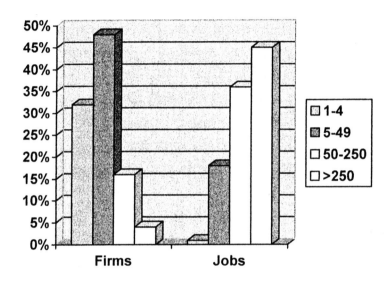

Figure 4. Chemical Industry Employment: Smaller Firms Account for Majority of Jobs (% of Totals by Firm Employment Size) Source: U.S. Bureau of Labor Statistics 2002

67% of all chemists surveyed worked for industry, vs. 25% in academia, 7% in government, and 1% self-employed (presumably consultants). These ratios are virtually identical to those of two years earlier, when they were first broken out." *(21)* As noted earlier, the nature of the survey cannot tell us reliably if or how overall total professional employment in industry has changed, only historical trends within its own parameters. While the 2005 survey showed improvement in unemployment, down to 3.1%, this appears to have come about almost entirely due to an increase in part-time employment; while full-time jobs remained "essentially unchanged at a historic low of 90.8%, down from 90.9%." *(22)*

At least equally significant is the reported stagnation in starting salaries during the past few years, strongly suggesting that the current supply of new chemistry graduates is either equal to or in excess of demand. The ACS salary survey states that while salaries showed a small increase, "in constant-dollar terms, however, median salaries for inexperienced new chemistry graduates remained depressed. When adjusted for inflation, the median salaries for 2003–04 graduates at all three degree levels were about 10% below the salaries received by chemists who had graduated three or four years earlier. As to employment, 38% of 2003–04 Ph.D. graduates found full-time permanent employment, up from 37% one year earlier. The gain for bachelor's graduates was also a nominal 1% – from 24% to 25%. For the smaller and more volatile master's class, the gain was bigger, from 41% to 48%. In 2000, the last really healthy employment year for chemists, a considerably higher 45% of Ph.D. graduates, 35% of bachelors, and 56% of masters reported that they had full-time permanent jobs upon graduation." *(23)*

The uncertain employment outlook, stagnation in salaries, and somewhat tarnished image of chemistry (the news media seem incapable of using the word "chemical" without joining it to the modifier "toxic") may have diminished the interest of U.S.-born college students in becoming chemistry professionals. It is also widely held that U.S. primary and secondary school educational curricula tend to discourage rather than encourage students to study science and mathematics. This latter situation reduces the number of U.S. born college-bound high school graduates who can even qualify to become science and engineering undergraduates without additional education.

For many decades time now, universities have maintained science and engineering enrollments as well as faculty staffing by attracting a sufficient number of non-U.S. nationals to offset the significant shortfall of U.S.-born students and professors. In effect, the U.S. has been "insourcing" about 40% of its scientific and engineering talent for a significant period of time. Following 9/11/01, non-U.S. student enrollments at the graduate level have dropped about 6%. (24) While it is quite likely that post-9/11 stricter U.S. visa requirements are a leading cause of this decline, they are by no means the only reason. Increased employment openings in the native countries of these graduates, e.g., China and India, also have much to do with this. Also, China has sharply increased its number of science universities, making it more attractive for native-born students to study at home. The U.S. will need to reform its scientific and engineering educational process if it wants to attract more U.S.-born students to make up for this shortfall.

Despite the currently sparse employment outlook for chemical industry professionals, demographic changes are very likely to reverse the picture in the very near future, in the form of retirement of the "Baby Boomer" generation (usually thought of as those born 1946-1964). These individuals are now reaching the age of 59½, when they become fully vested in 401(k) and other defined contribution retirement plans and may begin to draw benefits without restriction. This almost certainly will usher in a wave of retirements on a scale that the country has never experienced before. Companies urgently need to start planning now how to fill the vacancies that these retirements will create. Another problem is that the number of workers between the ages of 35 and 44 will shrink by 7% between now and 2012. This situation will demand a more flexible and adaptive approach to employment than the conventional one or employers may indeed find it unavoidable to relocate major manufacturing and R&D functions overseas to where qualified scientists and engineers are located. The subject of outsourcing R&D is discussed later in the chapter under the heading of "Is U.S. Chemical R&D in Decline?"

Since Americans today are in better health and enjoying longer lifespans, they are staying more active than has been typical in the past, including working full time or part time well past the age of 65. Companies that revamp their employment policies to retain and utilize this otherwise lost talent and experience will be able to compete more effectively in a globalized economy. (25) Those firms that do not will find their ability to compete greatly compromised. The situation is not limited to the difficult of recruiting from a diminished pool of chemical professionals – the ranks of industry management will be affected, too.

Of course, the chemical industry will not need to replace so many of those retiring Baby Boomers if it is forced to relocate production facilities overseas, as discussed earlier.

The Escalation of Competition

As mentioned earlier, business matters are not becoming easier for industry managers. Competition is growing keener all the time, and one of the leading reasons is that computers and the Internet have made it easier for both consumers and industrial buyers to obtain competing bids to supply their needs, almost instantaneously. Of course, it takes some time to qualify acceptable sources for chemical product and service offerings, but once this has been accomplished, it has become routine to solicit bids from suppliers in countries all over the world (an essential part of globalized trade – to buy as well as sell internationally). The bid-and-buy process has become both transparent and executable in real time, thanks to supply chain management practices. In many respects, the chemical industry has lead the way in this area through its adaptation of enterprise resource planning in the early 1990s and follow-on information technologies.

The catch phrase, "China Price," has come to symbolize the competitive nature of globalization, not only because Chinese firms typically offer the lowest prices, but also because qualified competitors wherever they are located in the world are also trying hard to match these bids. Since competition has become so much sharper than at any time previously, offshoring and outsourcing are now as much a survival tactic as they are a way to improve profitability. Companies must decide quickly what products and services they need to keep in-house and which ones they can more effectively and efficiently obtain from others. These are not one-time decisions, either, and must be reevaluated regularly. Prime examples of what should stay in-house indefinitely would be proprietary formulations and process technology. Additionally, firms must ensure that they have factored in all of the costs for offshoring, such as air freight for rush deliveries and/or additional inventory costs, bearing in mind that ocean shipping from Asia to the U.S. can take up to six weeks and that demand surges cannot always be forecast accurately. Quality issues need to be considered very carefully, as do the problems of communication between dissimilar business cultures.

Ultimately, global competition has been heightened most of all by sluggish domestic demand in Japan and the largest countries in Europe over the past several years. Annual GDP growth rates of less than 2% for some time now in these countries have meant that increasing numbers of manufacturers are finding significantly diminished domestic demand for their goods compared to the 1990s. This situation, in turn, has further intensified the drive for cost-cutting and placed additional pressure on manufacturing employment. The situation is what *Forbes Magazine* publisher, Rich Karlgaard, calls "the cheap revolution," i.e., goods are constantly becoming cheaper all around the world, which makes them more affordable for people, even in developing countries, but also causes rapid turnover as well as decline in global manufacturing employment.

Industry and the Environment

The chemical industry has had a weak environmental reputation in the eyes of the public, hurting its credibility despite its significant gains in reducing waste, discharges of pollutants, spills, and improving an already far better-than-average manufacturing plant safety record. Even though the American Chemical Council's Responsible Care program has broadly raised industry standards and compliance with safety and environmental controls, not every chemical company is a member of ACC, and some of these firms have had well-publicized pollution problems. Industry critics have seized upon this issue and want government- enforced controls to ensure that every company complies. Advertising and public relations campaigns to acquaint the public with the industry's improvements have met with only limited success. Unfortunately, the industry as a whole appears to be still viewed by the public as being no more environmentally conscious than its firms that are the least so. The extension of the Responsible Care program to non-ACC members in other countries (the "Global Charter") should be a positive step toward improving the world-wide chemical industry's performance, which will be the only sure way to improve the industry's reputation.

As a result, the industry's reputation has led to a federal and state regulatory climate with environmental laws that have more stick than carrot, making U.S. safety and environmental regulatory compliance costs greater than anywhere else in the world. This means that an investment dollar will buy less production capacity in the U.S. than it will offshore, regardless of the nationality of the investor. Therefore, international investors are more likely to build plants offshore where environmental regulation is often much more lax. Needless to say, this situation puts U.S. chemical industry jobs "at the margin," while not providing much, if any, environmental improvement from a global standpoint.

Perhaps a prime example of how the entire U.S. chemical industry has been haunted by the history of a single event is the tragic Union Carbide India Bhopal accident of 1984. Even though the disaster happened in India and was 49% owned (and 100% operated) by an independent Indian company, the resulting adverse publicity was no less intense than if the event had happened in the U.S. at a Union Carbide plant. The financial and legal fallout from the calamity eventually destroyed Union Carbide, which sold off parts of itself to survive before being finally acquired by Dow in 2001. The cause of the accident has never been resolved in the court of public opinion (Union Carbide's investigation showed it was sabotage by a disgruntled former worker, while tort lawsuit attorneys claimed it was due to inadequate safety procedures and management practices). The catastrophe claimed thousands of lives but one reason it is still revisited regularly even in the industry press is that compensation to victims or their families from the $470 million Union Carbide paid to the Indian government has been incredibly slow reaching them; only $325 million, less than 70%, had been disbursed as of January 2005. (*26*) Since Union Carbide no longer exists, tort attorneys have now turned to suing Dow, as its successor company, to put up still more money to compensate the victims. The Bhopal situation is then used by issue groups to tar the entire chemical industry as heartless and uncaring. Effectively, this has become a no-win situation for the chemical industry as a whole.

As mentioned earlier, EU environmental compliance costs have historically been less than those in the U.S., but this would change abruptly if the EU adopts the proposed REACH policy – Registration,

Evaluation, and Authorization of Chemicals – in its present form. Effectively, the policy would require extensive toxicological testing of every chemical manufactured. European chemical manufacturers see this proposal as so far-reaching and expensive, that its adoption in its existing form would put them at a disastrous disadvantage vis-à-vis non-EU competitors. A change in EU leaders has recently given industry groups hope that REACH will be softened by increasing the minimum annual production quantity that would fall under the REACH provisions, from one to ten metric tons. The U.S. Senate is considering a measure similar to REACH, known presently as the Lautenberg-Jeffords bill, but it seems unlikely that this will become law under the current political alignment.

Globalization & Regional Markets

While it was popular for business leaders 10-15 years ago to talk about entering Asian markets in Japan and the "Seven Tigers" (Singapore, Taiwan, Thailand, Malaysia, South Korea, Indonesia, and the former Crown Colony of Hong Kong), today China and India are very much "the" hot places to be. Indeed, this makes sense – the population of these two countries represents about one-third of the world total. Both China and India have a rapidly growing "consumer" class, currently estimated at a combined more than 250 million people – almost ninety percent of the *total* population of the U.S. The rates of economic growth in each of these countries are more than double those of the U.S. and EU rates, 8-10% vs. 4% in the U.S. and 1-2% in the major EU countries. With such great potential, the Western chemical industry simply cannot afford *not* to be in China and India, without risking that they will drop to second tier companies.

However, great opportunity is usually accompanied by great risks. Competition inside China is fierce – there are many more companies chasing smaller amounts of business than are usually found in Western countries. Local firms are encouraged by the government to focus on growth more than earnings (to increase employment). Furthermore, almost all of the larger companies are partially or wholly owned by the Chinese government, and government-owned companies make very formidable competitors. Foreign investors will find that the ability to repatriate earnings is severely limited except via exports. In fact, the

Chinese government uses this policy to "encourage" foreign businesses to reinvest earnings locally rather than paying dividends to overseas owners.

The Chinese government insists that companies who build facilities there bring with them the latest technology, particularly if they are joint ventures, but Chinese legal protection of intellectual property has proven to range from "uneven" to wholly inadequate (although chemical industry patents have fared better than others). Corruption, by Western definitions, is extensive. Pollution, both from industry and automobiles, is on a scale not found in the West for more than fifty years ago; environmental regulation is weak and sporadic. Additionally, the Chinese political situation holds significant potential for instability, a situation that has cropped up many times before in the 3000 year history of China. The rapidly growing income disparity between the coastal areas and the impoverished interior hold potential as a significant source of political unrest, as do the higher political aspirations of the newly-minted consumer class. If the Chinese government were to attempt taking over Taiwan by force, which it has threatened to do on a number of occasions, there would surely be a serious and extended disruption of its domestic economy, and for that matter, the economies of all its trading partners as well.

Other Asian countries besides China should not be overlooked by industry leaders planning to capitalize on overseas growth opportunities. Indeed, India offers much of the current high growth potential of China but with fewer risks. India has a more reliable legal system and a stable, functioning democracy. English is India's official second language, which has facilitated its growing and significant international service industry, as well as R&D centers for Western companies. India has the highest birthrate of any major economy in the world and, even though this rate is slowing, India's total population is on track to catch up with and pass China's total population within roughly two generations. (27) India's future population will also be more youthful than China's because China's state-enforced population control program (one child per family was instituted in 1979) is pushing it toward a much more mature population. By 2040, it is estimated that China will have a greater proportion of people over the age of 60 than the U.S. (which has the second highest birthrate of any major economy). Because of these demographics, India's economic development in the future is likely to be more stable than that of China. (28) Note that population projections for up to 20 years into the future are based on the existing generations continuing their current reproduction patterns, so that those for beyond 20 years are necessarily much more speculative.

Eastern European countries, particularly Poland, the Czech Republic, and Hungary, also offer significant future growth potential. By joining the European Community, they will be able to participate in the Euro-zone. Their lower labor costs make them a very attractive location for firms seeking to gain access to western European markets without paying customs duties. Russia remains much more of a conundrum. While it has a substantial population, current birth rates are actually negative (more abortions than live births) and the political climate has proven unpredictable in its attitude toward foreign investors.

Other less developed regions, such as South America and Africa, have been far slower to improve their economic climate for growth than most Asian countries. Most economists attribute this situation to political limitations rather than insufficient population growth or lack of capital. Even Mexico, linked directly through the North American Free Trade Association (NAFTA) to U.S. markets, has been so far unable to provide a sufficiently business-friendly political climate to be able to employ much of its population. The result has been emigration on a massive scale to the U.S. in order to find jobs. If Mexico were to remove these barriers to economic development, it is conceivable that its economic growth rate might well double, to the great benefit of both countries.

Influence of the Financial Community

Few chemistry professionals who are not involved in senior corporate management fully appreciate how critical the need for affordable capital is to the chemical industry or what hoops senior corporate management has to jump through in order to attract and hold the providers of that capital. The larger any corporation grows, the more it is likely to become dependent on attracting capital from the public and institutional investors to continue growing. The chemical industry is very capital intensive and therefore turned early in its history to the public as a source of funds. While nearly half of all Americans now own stocks, few manage or even own them directly. John Bogle, the pioneer founder of Vanguard, claims that America's 100 largest money managers now hold 58% of all stocks, primarily on behalf of mutual and pension funds. Chemical industry stocks holdings are effectively the same as those found in the rest of the stock market, with only 30% held by individual investors. (29)

Both individual and institutional investors are protected in principle by regulations promulgated by the Securities and Exchange Commission (SEC). Both groups also rely on the reports and forecasts of financial analysts who specialize in different stock market sectors and companies, keeping a close eye on their financial performance. However, these analysts generally tend to cater to institutional investors who, in turn, are most interested in financial results over the next year, and occasionally, the next two to three years. Further reinforcing the emphasis on relatively short-term results is the changing nature of corporate boards of directors. Pressure from watchdog groups as well as from the Securities and Exchange Commission (SEC) has lead to an increasing percentage of independent (non-management) board members in many publicly held companies. These directors often do not have chemical industry experience and are more accustomed to demanding short-term financial results.

Bear in mind that institutional investment managers are being judged by the pervasive demand for results – if the funds they manage do not yield at least average results over the course of their management contract, it will not be renewed. So the ultimate responsibility for the focus on short-term results must be laid at the feet of the public, who vote for the best results by switching their mutual fund and 401(k) and IRA investments to those companies whose recent financial track records are best. In due course, this applies to choosing banks and insurance companies, too, because the price of the services they offer is affected by the outcome of their investments in addition to their efficiency in delivering financial services. Pension fund and other large institutional managers may be somewhat more insulated from immediate judgment, but their clients read the stock market results as well and are equally demanding, even if over longer-term contract lengths.

It is an unfortunate fact of life that the chemical industry has been steadily falling out of favor with the investing public and financial analysts over the past 30 years. Perhaps the leading reason is that many investors and analysts evidently consider the chemical industry as a whole to have become mature; such investments are definitely not favored as "buy-and-hold" stocks. This has lead to the breakup and disappearance of a number of old names in the industry, e.g., American Cyanamid, Allied Chemical, etc. To add insult to injury, financial analysts have classed pharmaceutical companies as a separate stock market sector from the chemical industry, thereby isolating one of the fastest growing segments of our industry and ensuring that the remaining companies left in this narrower definition of the chemical sector are indeed more mature.

The investment policies of institutional investors and SEC regulations also unintentionally work to disfavor U.S. chemical companies. For example, if a fictional "Major Chemical Company" reported earnings that fell below expectations for several calendar quarters, it is likely that a number of the investors will sell their holdings, in order to reinvest in a more promising stock, driving down the price of the stock (more "sell orders than "buy" orders will cause the price of a stock to fall). Under these circumstances, an equally fictional "Big Pension Fund" would be likely to find that the percentage of its holdings in Major Chemical Company have now risen to a level that requires reporting the extent of these holdings to the SEC, and also now exceed its internal guide lines for the maximum percentage of the outstanding share of any firm that can be held. Not wanting this added administrative burden and needing to comply with its investment guidelines, Big Pension Fund decides to divest most or all of its holding in Major Chemical Company, further depressing the price, and making the stock of Major Chemical Company look increasingly risky for other holders. This multiplier effect on the volatility of "underperforming" stock prices, together with the lack of patience present-day investors have with cyclical stocks, have almost certainly caused the collective value of chemical stocks to shrink. Coupled with the emergence of the large "tech sector" (computer and Internet related companies) and the high favor that financial and other stock sectors now enjoy, chemical stocks have dropped from representing 8% of the S&P 500 in 1968 to 1% in 2001. *(30)*

All publicly-held companies face a quarterly challenge: continually increase quarterly earnings or fall from favor. Losing favor means a falling share price. A falling share price means becoming a take-over target, as well as putting employee stock options "under water." The basic problem for the chemical industry is that earnings are typically cyclical, but investors want constant growth. The demand for continuous earnings improvement has lead corporate management to a state of constant change that have generally included

- Consolidation via mergers and acquisitions;
- Divestiture of older, as well as smaller, business units;

- Restructuring e, e.g., personnel reductions and outsourcing;
- Shifting R&D focus away from new product discoveries toward shorter-term objectives, such as process improvement and customer service.

While these activities are usually presented as being in the long-term interests of the company and its investors, they are too often undertaken as short-term, quick-fix solutions to the need for ever-increasing quarterly earnings. One has the impression that too many senior corporate executives have become so involved in portfolio shuffling and restructuring, that they now have almost no involvement in what should be their primary focus, namely, growing existing businesses, and using R&D to introduce new products.

One part of this problem stems from overdoing the practice of moving managers from one position to another within the company to broaden their experience. While this is a desirable objective in itself, the time-in-place has been shrinking and is now as little as twelve to eighteen months in a number of larger firms for "fast track" managers. This is far too short a period for any professional to learn in depth the peculiarities of the business or technical area for which he or she is responsible and to undertake well-planned programs to grow that area. The opportunity to experience how business sectors perform during both economic expansion and contraction is also lost by such short tenure. Typically, it can take up to three years to develop, plan, execute, and demonstrate continuing significant growth for a business sector – by the end of eighteen months the jury would still be out. An exception might be where the lifespan of a product is less than eighteen months, but such instances are comparatively rare in the chemical industry. Consequently, the individual manager finds that the expedient thing to do is focus on cutting costs because this will show immediate results to his or her benefit when performance reviews are written. Then that individual will have moved on to the next assignment before it becomes evident that the immediate benefits do not have lasting effects – one cannot forever produce prosperity only by cutting costs. The business/technical area will eventually begin to fail due to lack of resources and, sadly, is then closed or divested to someone who is willing and able to restore those needed resources.

Many privately-held companies face difficulties in raising sufficient capital to grow. Huntsman Chemical has found a successful way to overcome this handicap:

- Remain privately controlled while financing acquisitions and expansion via joint venture or debt (which is often publicly-held in the form of bonds).

- Float public stock offerings when the market is high (but never ceding control), and buy back when the market is low. The proceeds of the stock offerings are used to pay down debt, part of which was issued previously to buy back stock. Unlike many financial investors who usually take companies public in order to pocket the proceeds, Huntsman uses this financial technique to strengthen its balance sheet and add to its flow of investment funds from depreciation and retained earnings.

The founder and chairman of Huntsman Chemical, Jon Huntsman, once said that the investing public doesn't understand commodity chemicals and lacks the patience or confidence (or both) to hold on to their investment in such companies through the cycles when earnings are particularly weak or lacking altogether. While he also thought that there could be a place in the portfolios of large, publicly held companies for cyclical businesses, it seems to be the perceived wisdom today that public investors will not tolerate corporate management hanging on to any business segments that do not show consistent earnings performance that is above average in all economic environments. Despite the efforts of publicly held chemical companies to cater to this apparent commandment, there has been discouragingly little evidence that these firms have then been rewarded by investors with greater willingness to buy and hold their stocks.

Huntsman is not the only highly successful, large chemical company that is privately held. Koch Industries, Intertech, JM Huber, Archer-Daniels-Midland also come to mind as firms that have found the secret to raising large amounts of capital without having to accept the problems of public ownership.

Duplicative or Differing Visions?

The senior executives of a number of publicly-held chemical companies appear to have followed much the same acquire-and-divest strategies over the past one to two decades. It is interesting to compare the courses taken by several of the major chemical companies and see how financial markets have viewed the results obtained. For purposes of this analysis, DuPont, Monsanto, and BASF will be compared. DuPont acquired oil and refining and then divested, entered pharmaceutical and again divested, spun off its elastomers into a joint venture and then pulled them back, divested basic fibers and some smaller specialties; it has kept most of the rest of its businesses, including agchem and related biotech. It has announced that it is now concentrating its corporate energies on the growth markets of Asia. During much of the same period, Monsanto spun off plastics, fibers, and basic chemicals acquired and divested pharmaceutical; it now concentrates on agchem and related biotech, very much on a global basis.

In contrast to the relatively high level of acquisition and divestiture activity by DuPont and Monsanto, BASF has followed a more conservative path. BASF, "The Chemical Company," has held perhaps the most unswerving business focus of large chemical companies, and consistently has been one of the three largest (in terms of sales) and best-performing (in terms of earnings) chemical companies in the world over the past decade. BASF's primary strategies have been to

- Hold fast to its *verbund* (integration) concept for over four decades, which features closely integrated manufacturing on a world scale at designated sites in Europe, North America, and Asia. BASF has now offered to become a "landlord" and extend the benefits of *verbund* to unrelated "tenant" companies at these plant sites.

- Stay committed to oil and petrochemicals, although not in exploration and development of oil.

- Divest infrequently and then only its smallest, least profitable units, with two important exceptions, which are cyclical large businesses:
    - Basell (polyolefins) – at the peak of its current sales/earnings cycle and highest valuation in 2005. While this has been a high-growth rate business, it has proven very difficult to generate earnings consistently. Price-cutting to maintain market share at any cost in down markets has long been an unpleasant and ingrained characteristic of the

polypropylene industry. BASF tried hard but unsuccessfully to make this business work via joint ventures: Basell was a joint venture with Shell, the result of both partners pooling their polyolefin businesses. BASF's polyethylene business had much earlier been put into a joint venture, Elenac, with Hoechst, but this partnership was dissolved in order to incorporate the business into Basell.

- o Nylon fibers via a portfolio swap with Honeywell for the latter's engineering plastics business in 2004. This move both took BASF out of a declining business and also markedly bolstered its North American nylon engineering plastics business. As a sign that BASF is committed to this latter business, it further strengthened its presence by acquiring of Ticona's nylon 66 compound businesses in the same year (Ticona had stopped polymerizing nylon 66 and limited itself to compounding in the previous year) and recently acquired the North American business of Lati, an Italian compounder whose business has mainly featured nylon compounds. These moves are parallel to those of DuPont and Monsanto insofar as all three have withdrawn from nylon fibers, but their moves differ in very important ways: Monsanto spun off its fiber *and* engineering plastics business, whereas DuPont kept the latter business segment, and BASF has been significantly bolstering its position in engineering plastics.

- Made China the major focus for future growth long before most competitors and, as a result, probably has the closest relationship with the Chinese government of any foreign chemical company.

It is beyond the scope or the purpose of this brief discussion to analyze which of these three companies has been more successful in running their business via the strategies they have chosen. Nevertheless, it is possible to observe what the sum of the assessments by institutional and individual investors in the stock market think of these companies' future prospects through buying and selling their shares.

One might assume that DuPont and Monsanto common shares would seem rather much the same to stock pickers, as they both appear to be following comparatively similar corporate strategies. However, the stock market shows that the combined judgment of investors results in very different valuations of these two companies. Figure 5 shows the quarterly closing price of Monsanto and DuPont common shares (adjusted for share splits) over the past five years vs. each other and the Dow Jones Industrial Average (DJIA). The figure shows that the three chemical companies have fluctuated within a relatively similar band, below the DJIA until mid-2003. From that point on, DuPont has been relatively flat, while BASF has moved modestly up and Monsanto has moved up strongly, exceeding the DJIA.

There are several possible explanations for these results, one being the obvious one that Monsanto is seen as finally having turned the corner with respect to improved growth and earnings, BASF also, but more modestly. DuPont's approach has not evidently found much favor and it has recently announced a share buy-back program. One explanation for these differences might be that Monsanto has a more narrowly focused business strategy and this is often favored by analysts and investors over a more diversified approach because the latter is hard to analyze and predict an outcome. BASF has fewer non-chemical business units than DuPont and therefore may also be viewed as having a more readily understood model than DuPont. However, not too much should be read into BASF's better performance because the majority of its shareholders are German institutions and individuals, who have a much more conservative buy-and-hold investment philosophy than the more speculative "day-trader" approach noticeable in the U.S.

Is U.S. Chemical R&D in Decline?

Available data on chemical industry R&D tend to be rather selective, usually only from large, publicly held firms. This makes analyses and conclusions only meaningful with respect to the sample, not to the overall picture. Some analysts use the number of patents issued or papers published to gauge R&D output, but these data do not differentiate between the value and significance of patents and papers and are therefore only modestly helpful. The aforementioned industrial R&D spending data are likely a more useful tool, as at least they tell us what the sum of decisions by large firms is on how much to invest prudently in future discoveries; trends over the past ten years are shown in Figure 6.

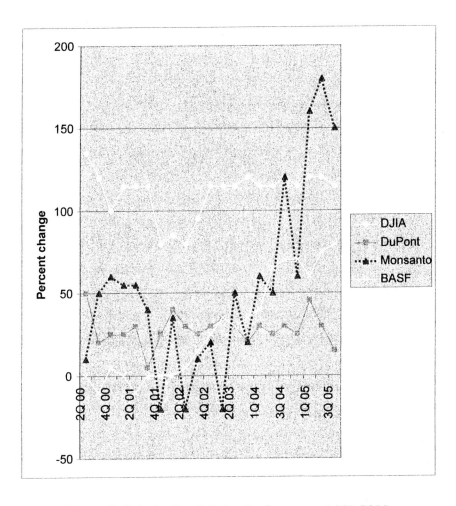

Figure 5. Relative Stock Price Performance, 2000-2005
Source: DuPont, Monsanto, BASF

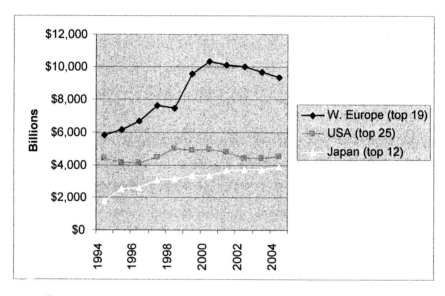

*Figure 6. Regional Chemical Industry R&D Spending Totals
Source: C&EN July 11, 2005*

U.S. domestic chemical industrial R&D spending (excluding pharmaceuticals) began declining in 1994, but has recovered in 2005 to the level it had in 2000. It is true that some R&D has been outsourced by companies, but, more often than not, this has been kept within the U.S. at contract research firms and universities. More recently, however, some large firms appear to be expanding their R&D activities in offshore locations rather than in the U.S. For example,

- DuPont has built and expanded technical centers in Asia – India, Singapore, China – in order to be close to what as it views as its highest potential growth markets. It does not appear to have reduced its U.S. R&D effort in the course of implementing this strategy.

- GE has doubled central R&D staff over the past eight years, but *all* of the new jobs have been located offshore, in India, China, and Germany; the latter two locations were built in the last three years. Note that Germany is at least as expensive as the U.S. in which to conduct R&D, so it cannot be said cost is the primary incentive underlying GE's decision to build this new offshore site (unless there were undisclosed German government financial incentives involved).

This is certainly not a bad thing – U.S. companies must be competitive in overseas markets or risk losing business to local and other international rivals. In words attributed to GE's former CEO Jack Welsh, companies need to plan globally but act locally.

Fortunately, there is solid evidence that the growth of U.S. industrial R&D spending offshore has been more than offset by the growth of overseas companies spending on their own U.S.-based R&D activities. Clearly, offshore companies see a benefit from maintaining R&D facilities in the U.S., whether it is in support of local markets and customers, or taking advantage of specific local technological skills and specialties. In 1997, the most recent data period, total U.S. industrial R&D spending overseas amounted to $14 billion, which is 11% of industrial R&D performed in the U.S. In the same year, overseas firms spent nearly $ 20 B on R&D inside the U.S., 15% of their total R&D expenditures. Industrial chemicals R&D spending by the U.S. subsidiaries and affiliates of foreign companies amounted to 20% of the total industrial chemicals R&D spending in the U.S. *(31)* Thus, R&D

funding in the U.S. has been receiving more from foreign sources than U.S. companies have been spending abroad. The data are too fragmentary and incomplete to be able to project any significant growth trends, unfortunately.

In Europe, German chemical industry R&D has also been declining, since 1987, but funding remains greater than in the U.S. as a percent of sales, 4.0% vs. 2.5% (these data include pharmaceutical companies). Unlike U.S. firms, German corporations that outsource R&D appear to favor contract research companies more often than universities. (*32, 33*) Data on overseas vs. in-country spending are unavailable.

Chemical industry R&D funding in the United Kingdom rose to 2.5% of sales in 1990 and has since tracked U.S. levels through 2003. Forty percent of UK chemical industry R&D is performed overseas, while overseas sources fund over 25% of R&D performed within the UK. (*34*) Thus, the UK has more net "outsourcing" of R&D to be concerned about than does the U.S.

In Asia, Japan is the largest player in the chemical industry. Japanese chemical R&D funding has grown faster than U.S. funding, increasing every year since 1980, and is now equal to Germany as a percent of sales. Again, these data include pharmaceutical companies. Interestingly, Japanese firms lead all others from overseas in the number of research facilities in the U.S. that they maintain. (*35*) As with German companies, overseas vs. in-country spending comparisons are unavailable.

In the author's opinion, U.S. non-pharmaceutical chemical firms as a whole appear to obtain the most productive return on their R&D expenditures vs. offshore competitors, in terms of industry growth, new products, and patents in use. European and Japanese chemical companies have long spent more on R&D than U.S. chemical companies but have not been able to make any headway closing the gap in terms of national aggregates of sales and earnings. The challenge for all regional chemical companies, therefore, is not merely to sustain or increase R&D expenditures, but to improve results.

What Might Be Done

Although it may appear obvious, it is worth repeating that every company needs to have a global strategy, whether it actually operates plants overseas or not. Otherwise management risks being overwhelmed by change and competition that seem to come out of nowhere. At the same time, it must be noted that not every company needs to have a subsidiary in China – it is a major opportunity but is also a high risk location. Much depends on a company's customers, as well as its technology and cost base, when making the decision about having a direct presence in China; the lower risk basis would be to partner with a Chinese firm, either through collaboration or a joint venture. After a period of several years in which to gain experience, then management can decide whether the time is right to take the plunge and establish a subsidiary.

When it comes to making earnings grow consistently in a highly competitive marketplace, management needs to keep in mind that quality improvement leads to cost reduction, not the reverse. A tunnel-vision focus on cost reduction alone can eventually put a company out of business by cutting its innovation funding to the point of ineffectiveness. Companies that have cut people and equipment costs to the bone are also vulnerable during growth spurts – they cannot ramp up production quickly (doing so invariably courts quality and cost problems) and risk losing market share to competitors.

Perhaps the most difficult question facing U.S. chemical company management today, particularly in small to medium size firms, is to identify which business sectors offer the best potential for future profitable growth in a globalized market. Although perhaps counterintuitive, the best earnings growth opportunities are likely to be found in those markets that are the *least* globalized, even though they may be only niche opportunities. These include sectors that are predominantly concentrated within a country or free trade agreement area, are to some extent regulated, and are technology intensive. Examples might include specialty packaging, on-site process services, and food additives. On the negative side, virtually all commodity products, particularly those with a relatively high labor component, such as textile fibers, will be under almost prohibitively greater offshore competitive pressure in the future.

Making the shift into specialty markets may also entail exiting those mature commodity businesses mentioned earlier, and sooner rather than later. Nevertheless, not every commodity business should be jettisoned. If the operation has established itself as either number one or two in the global market, has a technology advantage over competitors, and a reasonable expectation of continuing positive cash flow, then it should be kept. A greater challenge, but potentially a very rewarding one, would be to find ways to differentiate sectors within commodity businesses; in effect, turn commodities into high-volume specialties.

If the analysis shows that divestment is necessary to recoup invested funds while it is still possible, then the next question is how to replace the divested business. If a company does not have any major new products under development (and if not, why not?), then it will have to seek one or more existing businesses in the preferred new areas to acquire. A successful merger and acquisition (M&A) program, however, is not simply a matter of finding a target, making an offer, and signing a contract. The industry is replete with failed acquisitions, and these can be very costly in time, investment funds, and lost opportunity costs. In order to execute a successful M&A strategy, senior managers, especially in large companies, have to make some drastic changes in thinking.

The first and most important thing would be to acknowledge that the managers of the acquired company know their business far better than the managers of the acquiring company and emphatically do not need new senior managers from the acquiring company transplanted into the acquired company to run the operation. Nevertheless, transferring some good junior managers to gain experience in a new milieu would be very sound. Second is the need to understand and accept the fact that while earnings growth is more important than sales growth, the latter is a necessary component of improving earnings, provided it is accomplished on a sound basis, e.g., do not sell products at prices below full cost in order to fill plant capacity. Third, manage for the long term and do not cut R&D during recessions, as doing so greatly harms future earnings growth and loses ground to competitors, which will be extraordinarily difficult to regain. Fourth, develop some tolerance for entrepreneurial risk-taking. Failure, as long as it is not of "bet-the-company" magnitude, must be viewed as part of learning how to innovate through experience. The outcome of research is never assured success or it would not be research, only applied development. The basic lessons learned from such works as Peters' and Waterman's 1980 book *Search for Excellence* about how to succeed in business, e.g., "the company forgets what made it successful in the first place, which was usually a culture that encouraged action, experiments, repeated tries," are every bit as true today as they were 25 years ago.

As a less costly alternative to acquisitions, joint ventures can be attractive. Joint ventures spread risk as well as investment. The most successful joint ventures are those between companies of relatively equal size and similar business cultures; they are also usually limited to production and process R&D. If each partner markets its share of the joint venture's output, most antitrust concerns can be put to rest and financial accounting is simpler than if the joint venture also markets the output. The downside of a joint venture is that staffing it exclusively with personnel from the parent firms can lead to conflict; again, this is minimized if the joint venture is limited to manufacturing products for each parent to sell. [37]

Lastly, the U.S. government needs to make some political reforms that will ensure manufacturing is not forced to labor under significantly greater cost burdens than those of our international competitors. It is of critical importance to the U.S. chemical industry that additional domestic sources of natural gas are developed and that pipeline transport is facilitated, not hindered. Unless Congress takes meaningful and immediate action to increase domestic natural gas supplies, much of the U.S. basic chemical industry will almost certainly be forced overseas.

Leaders in both federal and state governments need to acknowledge that a healthy manufacturing sector is necessary to the economic well-being of the U.S. – an "all-service" economy that some actually admit to wishing for, does not, and cannot lead to sustained national wealth generation.

Conclusions

Press reports about the movement of U.S. manufacturing offshore turn out to be purely anecdotal and selective, and certainly not representative of what is actually taking place. The real situation may be more clearly perceived from examining government statistics that show the country's manufacturing sector *per se* continuing to grow strongly, following the recession of 2000-2003. That being said, it is indeed true that both U.S. and total global manufacturing employment has been and is continuing to decline. The cause, however, is not the result of "offshoring" jobs, but rather continuing productivity improvements, and this has been part of a truly global phenomenon that has been taking place for more than thirty years.

Some sectors of the chemical industry appear to have matured in the past five to ten years, making earnings growth above GDP rates for these areas possible only through productivity gains and consolidation. Government policies have reduced industry competitiveness vs. overseas competitors in the areas of safety and environmental compliance costs, mandated employee benefits, tax rates, tort costs, and energy policies, with little relief likely in the latter situation before 2008 at the earliest.

Chemical industry professional employment in the U.S. appears to be weak to stagnant despite corporate economic growth during the current growth stage of the economy. The absence of real growth in entry pay scales appears to confirm this. However, the retirement of the "Baby Boomer" generation now getting underway will open up significant employment opportunities over the coming twenty years or more. The degree to which these openings will be filled by U.S.-educated scientists and engineers will depend largely on improvements being made in U.S. secondary school science and mathematics educational programs to attract and qualify students who will aspire to these professions.

The U.S. chemical industry's long-held cost advantage of olefin-based products will shift to Middle East by 2008. This prospect has already resulted in a shift of new capacity investment by many U.S. producers of these products to overseas locations. While the new Middle East production is targeted to be sold into growing Asian markets, the net effect will be to displace Western-made products.

The chemical industry has acquired a reputation for being the source of dangerous materials that despoil, pollute, poison, and explode. This disreputable perception is continually fed by sensationalist stories in the media and by environmental activists; efforts by the industry to improve its image have met with only limited success. Changing this perception requires education and performance, not press releases. Until the industry can make significant progress in recovering public good will, it will be subject to stringent and costly regulation that is not evenly applied around the world. Thus, chemical companies in the developing nations will have a cost advantage over those in North America, Europe, and Japan.

While a number of publicly held corporations have responded to investor concerns by almost constant portfolio shuffling and restructuring, sustained long-term improvement in financial results and consequent investor interest have proven highly elusive. Huntsman Chemical stands out from other large firms by keeping its stock under private control, using publicly held debt to fund expansion and acquisitions, together with private placement of common stock that does not exceed a minority interest position.

The chemical industry is responding to the challenge of globalization but each firm's management needs to find its own optimal solutions and avoid the groupthink that has marred the performance of many, including some old name companies in the past. In particular, it would seem to be a more effective application of management effort to concentrate on growing the firm's business than depending on acquisitions and divestitures to improve profitability on a sustained, long term basis.

Chemical industry R&D suffered during the recession of 2000-2002 as companies cut back heavily to avoid or minimize financial losses, but in 2004, finally returned to the same level as 1999 – a five year setback. Corporate executives are still deciding how to get more out of their R&D expenditures, but seem to be converging on a focus on serving customers locally, particularly in China and India. Cost- and time-effective R&D is crucial to producing new products and processes that generate growing corporate profitability. U.S. R&D appears to be more cost-effective than that conducted in other countries and overseas companies recognize this by conducting a significant portion of their R&D in the U.S.

Firms in the chemical industry need to update their analyses of the characteristics of the businesses they are in. Then they need to emphasize more profitable, specialized business sectors than irredeemably commoditized ones in order to grow earnings on a sustainable basis in the future. While acquisition of desirable businesses is still the fastest way to enter new products and markets, the industry's track record has been mixed at best, and large companies need to adapt their thinking to a more entrepreneurial mold before they will succeed (such as Dow Chemical's "intrapreneuring").

Finally, the U.S. government urgently needs to reform the cost burdens it has placed on manufacturing that greatly hinder its ability to compete on an equal basis with our largest trading partners. The government must also resolve promptly its contradictory energy policies that are rapidly making entire large sectors of manufacturing, particularly the chemical industry, uncompetitive in the global marketplace. This serious problem has yet to receive sufficient sustained attention in the U.S. Congress by either party.

References

1. *The Chemical Industry at the Millennium – Maturity, Restructuring, and Globalization;* Editor, P.H. Spitz; Chemical Heritage Press, Philadelphia, PA, 2003

2. Carson, J. B., *US Economic and Investment Perspectives – Manufacturing Payrolls Declining Globally: The Untold Story;* US Weekly Economic Update (Alliance Capital Partners, New York, NY), October 10, 2003
3. *Ibid.*
4. *Bureau of Economic Analysis National Economic Accounts,* URL http://www.bea.gov/dn2/iedguide.htm#gpo
5. *US Census Bureau 1997 Economic Census Subject Series,* URL http://census.gov/
6. *Bureau of Economic Analysis National Economic Accounts,* URL http://www.bea.gov/bea/dn/nipaweb/tableview.asp
7. *Central Intelligence Agency World Factbook,* July 2005, URL http://cia.gov/publications/index/html
8. Leonard, J., *How Structural Costs Imposed on US Manufacturers Harm Workers and Threaten Competitiveness,* White Paper, National Association of Manufacturers/MAPPI, Washington, DC, December 2003
9. Forbes, K. J., *US Manufacturing: Challenges and Recommendations,* National Association of Business Executives Washington Economic Policy Conference, March 25, 2004
10. *Ibid.*
11. Young, I., *China: Revaluation Makes History, Chem. Week,* August 24/31, 2005, p. 25
12. Bhidé, A. and Phelps, E., *Classical Theory vs. the Real World, The Wall Street Journal,* Monday, July 25, 2005, p A14
13. Browning, E.S., *Yuan Move Might Stir Big Ripples, The Wall Street Journal,* July 25, 2005, pp C1,4
14. *US Census Bureau 1997 Economic Census Subject Series,* URL http://census.gov/
15. Alperowicz, N., *Three More Ethylene Complexes in Qatar, Chem. Week,* January 26, 2005, p. 17
16. Alperowicz, N., *Iran Seeks a Top Spot in Petrochemicals, Chem. Week,* June 29/July 6, 2005, p. 39-41
17. Alperowicz, N, *Total Eyes PP Line in US; Raises Capacity in Europe, Chem. Week,* August 10, 2005, p.
18. *US Bureau of Labor Statistics [Monthly] Employment from BLS Household and Payroll Surveys,* URL http://www.bls.gov/ers/home.htm

19. *US Bureau of Labor Statistics Industry Studies,* URL www.bls.gov/oco/cgs008.htm
20. Heylin, M., *Employment and Salary Survey, Chem. Eng. News,* August 16, 2004, Vol. 82 Nr. 33, pp. 26-34
21. Heylin, M., *Employment and Salary Survey, Chem. Eng. News,* August 5, 2002, Vol. 80 Nr. 31, pp. 37-44
22. Heylin, M., *Employment and Salary Survey, Chem. Eng. News,* August 1, 2005, Vol. 83, Nr. 31, pp.41-51
23. Heylin, M., *Class of 2004 Starting Salaries, Chem. Eng. News,* April 18, 2005, Vol. 83 Nr. 16, pp. 51-55
24. Regets, M., *The (Continuing) Internationalization of the Chemistry Labor Force,* presented at the American Chemical Society National Meeting, San Diego, CA, March 13-17, 2005
25. Semple, L., *The Changing Workforce in the Chemical Industry,* presented at the American Chemical Society National Meeting, San Diego, CA, March 13-17, 2005
26. Tremblay, J.-F., *Bhopal Today, Chem. Eng. News,* January 24, 2005, Vol. 83 Nr. 4, pp. 28-31
27. *2005 World Population Data Sheet,* Population Reference Bureau, URL www.prb.org/pdf05/05WorldDataSheet_Eng.pdf
28. Fritz, M., *U.S. Birth Rates Remain High, The Wall Street Journal,* August 23, 2005, p. A2
29. Roberts, J., *And Then There Was One? Consolidation to Remain Major Industry Theme,* Merrill Lynch Equity Research Report, March 6, 2001
30. *Ibid.*
31. Dalton, D. H. and Serapio, M. G., Jr., *Globalizing Industrial Research and Development,* US Department of Commerce, Office of Technology Policy, Washington, DC, September 1999, pp 11, 33
32. Roos, U., *Research and Technology Policy, R&T Note No. 010 04 Date 9 March 2004, "Stagnation in German R&D Expenditure,"* British Embassy, Berlin, Germany
33. *FuE – Datenreport 2003/2004, Forschung und Entwicklung in der Wirtschatf, Bericht über der FuE Erhebungen 2001 und 2002,* Stifterverband für Deutsche Wissenschaft, Essen, Germany
34. *Research and Development in UK Businesses, 2003,* National Statistics, Newport, UK 2005

35. Dalton and Serapio, *op.cit.*
36. Anon., *Where's the Largest Market in a Decade?, Plastics Daily News,* August 17, 2005
37. Jones, R. F., *Strategic Management for the Plastics Industry,* CRC Press, Boca Raton, FL, 2004, pp 115, 122-123, 146

Chapter 2

Research and Development in the Pharmaceutical Industry and Investment in Innovation

Susan Wollowitz[1] and Faiz Kermani[2]

[1] Wollowitz Associates LLC, 455 Moraga Road, Suite C, Moraga, CA 94556 (email: sue@wollowitz.com)
[2] Chiltern International Limited, 171 Bath Road, Slough SL1 4AA, United Kingdom (email: faiz.kermani@global.chiltern.com)

Abstract

Constraints with the pharmaceutical industry are forcing a rethinking of where and how to place the industry's investment in R&D. Factors include market pressure in the existing major markets, return on current R&D investments, new options in carrying out R&D activities, and where innovation is occurring, among others. In this review we look at the major influences to creating an innovative R&D community and the current trends in R&D investment.

Introduction

The pharmaceutical industry, or biopharmaceutical industry if we are to include biologicals, represents the largest segment of the non-commodity chemical industry. Its unique features often require it to be considered as a complete separate industry group. One of the most significant features is the considerable amounts of revenue that the biopharmaceutical industry invests back into R&D relative to other technical industries (such as the electronics, communications, and aerospace) in the technology sector (1). What is less recognized is that the pharmaceutical industry is only one of the key investors in pharmaceutical R&D. There are a number of other private and public stakeholders in both the process and the outcome of pharmaceutical innovation that are driven to invest for different reasons. In addition there are forces that contribute to the downgrading of pharmaceutical research when the goals of the stakeholders come into conflict. The relative influence of these stakeholders and pressures vary globally, even among developed countries, leading to significant migrations and changes in the pharmaceutical research arena.

This chapter will discuss the various stakeholders, starting with the industry itself, the approximate size of their investments and the intent of their participation. It will also discuss some of the political and cultural constraints. It will then present the dynamics of the R&D investment environment in the U.S., Europe, and to a lesser extent, Japan. Since the field is in continuous flux, a current snapshot of a few of the newer centers of R&D activity in Asia will also be provided.

It should be mentioned that costs presented here are approximations only. While each organization publishes investment numbers, they often encompass a variety of activities of which only some are directly related to pharmaceutical R&D. We have not tried to interpret these ourselves but relied upon either the primary numbers or interpretations carried out by industry analysts some discrepancies are therefore inevitable. Unless noted otherwise, any direct cost comparisons come from sources using similar evaluation methods.

Innovation is the life blood of the pharmaceutical industry

In 2000, CMR International estimated that, on a global basis, pharmaceutical companies invested U.S.$58 billion in R&D, an increase of 121% since 1990 (2). In 2004, the top 20 global pharmaceutical companies alone, accounting for about 50% of pharmaceutical sales world wide, reported that they spent U.S.$58 billion in R&D (3), see Table I. This represented on average 19% of revenues, a far cry from the 3.5 to 4.5% in revenues that the rest of the chemical industry invests back into research (4). Even taking into account that only 30-40% of pharmaceutical R&D is spent on basic chemistry and life sciences related research and development, the other greater part going to clinical studies, the difference with the rest of the chemical industry is still remarkable. For manufacturers of biological therapeutics, the top 10 companies spent an average of 25% of revenues on R&D (5).

Table I. R&D Investment for Top Global Pharmaceutical Companies

Rank	Company	Revenue ($B)	R&D as % Revenue	Home Country
1	Pfizer	46.1	17%	US
2	GlaxoSmithKline	31.4	17%	UK
3	Sanofi-Aventis	29.6	17%	France
4	Johnson & Johnson	22.1	24%	US
5	Merck	21.5	19%	US
6	AstraZeneca	21.4	18%	UK
7	Novartis	18.5	23%	Switzerland
8	Bristol Myers Squibb	15.5	16%	US
9	Roche	13.8	30%	Switzerland
10	Lilly	13.1	21%	US
13	Takeda	8.5	15%	Japan
19	Sankyo	4.2	19%	Japan

Pharmaceutical companies are traditionally thought of as developing and selling small molecule therapeutics. Today they also sell biological products, often through partnerships and licensing agreements. Biological products are made by a smaller group of manufacturers; those that sell directly into the marketplace have a smaller portfolio and lower revenues as shown in Table II. Thus they are typically segmented out for financial comparisons, especially because up until recently biologicals were impervious to competition from generics. This allows them a very different R&D strategy although some of them are branching into small molecules and muddying the picture.

However, we must consider pharmaceutical R&D as covering both of these groups, sometimes called biopharma. The narrower term "biotech" has many different and confusing definitions. In the pharmaceutical industry it covers all enterprises that carry out R&D, except for the large traditional pharmaceutical companies. In this sense, biotech can include both the largest biologicals manufacturers and the small five-person start-ups working on a new screen for small molecules. This distorts the financial picture of this group, but only a very small percent of biotech firms sell directly into the consumer marketplace. In this chapter, the term "biotech" refers to those enterprises that feed into the larger biopharma companies.

Table II. R&D Investment for Top Biotech Companies

Rank	Company	Revenue ($B)	R&D as % Revenue	Home Country
1	Amgen	10.0	2028	US
2	Genentech	3.7	985	US
3	Serono	2.2	595	Switzerland
4	Biogen	2.1	685	US
5	Genzyme	1.5	391	US
6	Gilead	1.2	224	US
7	Medimmune	1.1	327	US
8	Chiron	0.9	402	US
9	Millenium	0.3	403	US
10	Intermune	0.1	81	US

From a government and investment perspective, the biotechnology sector also includes medical devices, diagnostics, and agricultural products. These different uses of the terms don't affect the discussion significantly, but it does mean that numbers and statistics provided from different sources do not always agree.

What is common to the wide range of enterprises that contribute to pharmaceutical sector productivity is the continuing commitment to research, even in the face of difficult economic circumstances around the world. The most obvious cause for this enormous investment is the relatively short effective patent life of pharmaceuticals, 11-12 years, vs. 18.5 years for other technical based products (6). These "ethical" drugs (those still under patent to the innovating company) have a substantially higher economic value than generics. Ethical drugs and biologicals constituted 80% of sales worldwide in 2003 [BCC 2004], while generics constituted 6%, the rest being OTC products. The high value of ethicals is underscored by the fact that while generics bring in a small portion of the revenue, more than 50% of prescriptions in the U.S. were written for generic products in 2004 (7). The large value differential and the rapid conversion of drugs to generic status push the pharmaceutical industry to constantly seek new drugs, and in particular, new ways of seeking, selecting, and developing drugs.

In addition, in the last few years there has been a gradual decrease in the number of totally new drugs that have been approved world wide, thus leading to questions about the innovative process in pharmaceutical companies as well as to the rapidly rising costs of R&D itself.

Thus pharmaceutical companies are faced with the dilemma of apparent run-away R&D costs, and an ever-narrowing operating margin in which to recoup their costs. The solutions to this require both changes in the R&D environment and the way that companies can achieve a return on their investment.

Pharmaceutical companies have attacked the problem from both ends. We mention briefly here that efforts to expand markets, reduce price controls, extend product lifecycles, and to shrink time to market, regulatory burden, and follow-on studies are key strategies for the pharmaceutical industry. The focus of this discussion however, is on the complementary issue of pharmaceutical R&D itself.

The pharmaceutical industry has also recognized that they must carry out R&D more effectively, or in essence, innovate the R&D process itself. A tremendous amount of effort has gone into examining, among other things, all aspects of how targets are selected, how compounds are made and screened, how information is collected, analyzed and communicated, why and how compounds fail in late stage development, and how to foster the innovative environment required to continue the process (8, 9). The innovation of innovation itself is accelerating. For example, specialized start-up companies have focused on expanding the range of drugs available to treat diseases and improving the efficacy of drug discovery (so-called platform technologies) which large pharmaceutical companies have obtained through mergers, acquisitions, alliances and licensing (*10*).

Large pharmaceutical companies and management firms have moved the concepts of portfolio management back into the R&D stages, with more detailed analysis of requirements for successful outcomes, responding earlier to failure modes, increasing flexibility to react to new technology, and enhancing the efficiency of their organizations (*11*). These efforts are research unto themselves with successes and failures, or require the development of additional technology or concepts before they are eventually successful. However, understanding and responding to the factors that slow R&D down can provide substantial savings benefits to the industry in their R&D operations.

Public health benefits from innovation

Public health is of course the key benefit from the existence of the pharmaceutical industry. Companies in the major drug development regions of the world, the U.S., Europe and Japan, have produced products to cover a range of therapeutic areas. Over the last 30 years these pharmaceutical regions have been responsible for the majority of the 1,400 new molecular entities (NMEs) launched as human therapeutics, which have made major contributions to improvement in healthcare (*2*).

Life expectancy has certainly been impacted by improved healthcare. For example, children born in 2000 can be expected to live nine years longer than if they had been born in 1960 (*12, 13*) according to a 2004 analysis of the 30 OECD countries by the Organization of Economic Co-operation for Development (OECD). The contribution of the pharmaceutical industry's products to this improved outlook for life expectancy is often overlooked, but vaccines are one area which clearly demonstrates the profound influence that its activities have had for the population.

A U.S. study found that pharmaceuticals accounted for only 9.4% of the total U.S.$1.3 trillion spent on healthcare in 2000 (*14*). Many new treatments aim to modify the diseases being targeted rather than treating only the symptoms and thus they can remove the need for expensive, lengthy stays in a hospital.

Several studies show that when effective medicines are used properly, early intervention in treating diseases can counteract some of the draining effect disease has on a country's economy. For example, chronic illnesses such as heart disease, cancer, and diabetes account for 75% of the expenditure on health care each year in the U.S. (*15*). The cost of effective drugs to treat these disorders and prevent them from worsening is considerably less. Recent studies have suggested that newer drugs lower mortality, reduce hospital admissions, patient visits, and days away from work relative to older drugs (*16*). The German

Association of Research-based Pharmaceutical Companies has cited figures suggesting that, in their country at least, for every €1.00 spent on a cholesterol-lowering drug such as a statin, €3.50 can be saved in hospital costs *(17)*.

Public health is coupled with convenience of treatment more than recognized. The development of more convenient drugs and formulations (i.e., once-a-day extended release products, or those with reduced side effects) is often denigrated as non-innovative, but R&D in this area can also have a substantive impact on public health. Patient compliance with pharmaceutical interventions is poor. In developed countries it is found that 40-70% of patients comply with recommended treatments for chronic diseases *(18)*. In the United States, where most health care consumers have little appreciation of product costs, yet substantial control of product selection, the drive to more convenient and palatable drugs is certainly largest. But the resulting benefits to increased patient compliance, patient health and thus the public health overall underscore the value that even these seemly minor innovations can have.

It is thus in the interest of government, pan-government, and non-government organizations to support the development of new drugs and formulations to reduce the economic burden of ill health on individuals and on the organizations themselves.

Regions want pharmaceutical R&D

A third group of stakeholders in the quest for innovation are regions that benefit from the existence of R&D and production facilities in their locales. In the U.S. and countries such as the UK in Europe, a little over 0.1% of the population is employed in the pharmaceutical industry, of which about a third is involved in R&D *(2, 19)*.

A high level of competition and innovation characterizes the pharmaceutical and biotech sectors. Together with advances in technology and improvements in their processes, these two sectors also rely heavily on the presence of talented staff to operate effectively. These highly paid professionals with favorable purchasing power, increase the tax base, and support an increase in cultural activities. Large R&D centers spawn numerous smaller R&D enterprises which further perpetuate and grow the economic vitality of the area. The success of a dynamic pharmaceutical R&D environment in a locale or region is a sign of a vital economy, giving confidence to an ever widening number of related businesses and resulting job opportunities.

Cities, states, national governments and economic regions all value such enterprises. Their investment in stabilizing and/or attracting R&D through direct financial input, or more subtly through favorable legislative and policy decisions has clearly had an impact on the health of regional R&D environments, as may be seen for example in the "hotbeds" of California and Massachusetts in the U.S., and more recently in creation of the Biopolis in Singapore. In contrast, reduction in general and specific R&D activities in a region have often been related to unfavorable political and legal climates, Germany being a key example as discussed later.

Venture capitalists and start-up enterprises

The over 5000 small and emerging biotech and pharmaceutical companies clumped together in regions throughout the world might be considered a tool of the industry rather than a driver. But in fact the relationship is symbiotic. Most such enterprises today do not seek to compete with, or become a fully vertically integrated pharmaceutical company, but rather to develop intellectual property, unique skills or early stage compounds that have high value to the large firms. Close to 30% of products under development today by "big pharma" are partnered with, licensed from, or based on proprietary discovery tools from small "biotech" companies. According to an Ernst and Young report, between 400 and 500 new alliances are made annually worldwide between large pharmaceutical and small "biotech" companies *(20)*. Thus, the goal of most biotech companies is not to sell a pharmaceutical product to consumers, but rather to sell opportunities to the traditional pharmaceutical companies, albeit with the hope of royalties from a successful product. Outsourcing of pharmaceutical R&D was worth about $9.3 billion globally in 2001 and by 2010 may reach $36 billion *[21]*.

Venture capitalists, typically investing in entrepreneurial start-ups that are local to them, provide a benefit to pharmaceutical R&D by fostering early stage research, and turn their successful endeavors back into the local economy. The vitality of this venture capitalist – entrepreneur pairing is strongly coupled to the willingness of the regional financial community to tolerate and encourage high risk enterprises.

Constraints on Innovation: Cost Containment – Public Health and Public Benefits Collide

There are many pressures that negatively impact the momentum of pharmaceutical R&D, especially on a regional basis. Certainly the rising cost of healthcare is one of them. As countries struggle with rising healthcare costs, placing pressure on pharmaceutical companies to reduce their prices has often been a favored policy. These cost containment policies have predominantly occurred in Europe and Japan, much to the displeasure of pharmaceutical companies, but are a growing feature of government policy worldwide. As will be seen this has important implications for the investment policies of major pharmaceutical companies, especially in R&D.

The problem for many governments is that the increase in healthcare expenditure has caught them off guard and so they have been slow to prepare in advance. Uncertainties in national economies combined with a growing elderly p opulation a nd f alling b irth r ates p laces a s train o n funding for public healthcare (see Figure 1) *(22)*. A range of measures are being used in different countries to control healthcare spending, but none of these approaches have been accepted as a universal solution, appropriately balancing short term need with a long term benefit.

Cost containment measures provide fewer incentives for companies in the healthcare sector to invest in new technologies. Pharmaceutical companies, pressured by the soaring costs of bringing a new drug to the market want relief to sell into a free market and are dissatisfied with many governments' current cost containment policies.

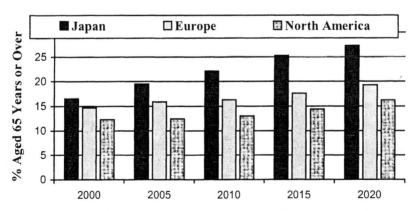

Figure 1. Global population projections: The impact of aging

Intellectual Property Rights: When Innovation Cannot Be Turned Into Profit

The huge cost of developing pharmaceuticals and in particular demonstrating their efficacy to regulatory authorities makes ownership of discoveries critical to the present paradigm of "ethical" pharmaceutical companies. They are highly protective of patents and reluctant to engage in regions where their patent rights are not recognized and where trade secrets are not well contained by employees. In the past, this concern has contained the pharmaceutical R&D within the developed countries that mutually respect and prosecute patent rights. Much of Asia, including India and China maintained differing views on patent issues that essentially wrote them off the map for establishment of R&D enterprises by large pharmaceutical companies. Since the somewhat controversial Trade Related Aspects of Intellectual Property Rights (TRIPS) agreement has come into effect, establishing almost global intellectual property rights, to new technology sectors, technology innovation enterprises have been more comfortable moving into these regions.

Effect of Globalization on R&D Activity

The pharmaceutical business is a globalized business and capitalizes on this for both increasing their markets and operating more efficient and cost-effective R&D efforts. Many of the current top 10 pharmaceutical companies in the world are mergers of companies from different countries and regions, or have acquired significant R&D and production facilities from outside their country of origin. Coupled with the ease of communication and the gradual harmonization of product and development standards, this has allowed the industry to select sites for manufacture and R&D that are in their best interests on a global basis. It has allowed them to react more quickly to government policies, economic climates, and innovative opportunities around the world. Figure 2 shows the split of pharmaceutical R&D expenditures in the key markets. The significant shift to outsourcing of these activities has also increased the flexibility of the industry to move about to best suit their economic needs on a relatively short term basis.

Nevertheless, the industry has shown a penchant for maintaining R&D facilities in their best market regions. Part of the motivation is the strong coupling of a strong talent pool and financial climate found in a population willing to pay for high valued pharmaceuticals. But in addition, the R&D facilities are positioned in regions where clinician thought leaders can engage in the development of innovative products and can champion them for their patients. This is the case regardless of the fact that large clinical trials themselves do not have to occur in the same countries or regions as the R&D activities.

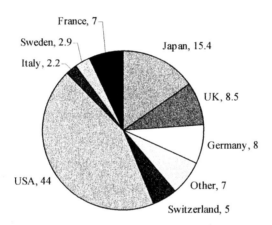

Source: EFPIA, PhRMA, JPMA, ABPI 2002

Figure 2. Split of Pharmaceutical R&D Spending in Key Global Markets

But the new world economy has altered both the markets and the opportunities for R&D options around the world. The rise of a middle class in Asia, in particular India and China, has therefore been of increasing interest to pharmaceutical companies that discover, develop and sell high valued pharmaceuticals. No region can control the balance within their borders of GDP-draining healthcare costs and GDP-increasing innovative R&D activities without recognizing and reacting to the global playing field.

Trans-Oceanic Dynamics

The U.S.A., Japan, and Europe are the key markets for the international pharmaceutical industry and make up around 80% of global pharmaceutical sales *(2)*. However, there are major differences in the financial policies, government roles, and cultural expectations of these markets which have affected the investment efforts and influenced those of the industries in the regions.

While the United States and the European Union are approximately the same size in population, the U.S. is a single country and Europe is a collective of many nations each operating their own healthcare system, and with different cultural priorities and financial policies. In the U.S. there is a single major health care policy, regulatory approval process, and financial setting. For the pharmaceutical industry, the U.S. is a single market where goods can be moved freely from one state to another, and the stock markets and capital investment are national. The shared primary language and general culture allows companies to attract and move talented personnel around the country with relative ease compared to moving personnel around Europe.

On the other hand, the U.S. is not wholly a single monolith as it impacts the pharmaceutical sector's R&D needs. The individual states vary considerably in the culture, economic and educational status and state and local governments play a major role in the economic vitality within their borders. However, in both Europe and the United States, the balance of power, policy making and economic investment lies with the national governments themselves.

Japan is the third significant ethical pharmaceutical market, with two pharmaceutical companies representing 4% of total revenue for the top 20 companies in the world. The market is as homogenous as that in the U.S., perhaps more so, but has a population and market size about half that in the U.S. In addition, it has been slow to freely interact with the larger pharmaceutical R&D community, missing some of the opportunities and risks of the borderless access to markets and human capital.

Over the last decade there have been major differences in the manner that the healthcare systems in these countries have developed and this has affected the way in which pharmaceutical companies invest in R&D in these regions. The pharmaceutical industry believes that its ability to discover and develop innovative new drugs depends on the competitive nature of the markets in which they operate and the availability of scientific talent.

These trends in R&D investment were highlighted at the 2000 Global Competitiveness of the Pharmaceutical Industry Symposium convened by the Directorate General Enterprise . The main finding of the report was that, as a whole, Europe was lagging behind the U.S.A. in its ability to generate and sustain the innovation processes necessary for pharmaceutical R&D. It noted that the situation needed urgent attention as R&D was becoming "increasingly expensive and organizationally complex."

An analysis of employment levels in the European pharmaceutical industry is revealing. Throughout the 1980s, employment grew steadily, at an average annual rate of just over 2% (representing about 10,000 new jobs each year), a rate that far outstripped that in other European manufacturing industries *(2)*. In 1994 this growth rate was suddenly reversed with 13,500 job losses (2.6% of the total employment). The industry has linked this sudden decline with healthcare reform and cost containment measures imposed by governments in certain European countries. For example, Italy's cost-cutting measures were estimated to have resulted in 7,000 job losses - a 10% reduction in total employment between 1992 and 1995.

The single factor of price controls has been blamed as the cause of the visible change in the pharmaceutical industry's view of Europe as a whole, something that has worried European politicians and industry experts alike (Figure 3). According to figures from the European Federation of Pharmaceutical Industry Associations (EFPIA), in 1990, pharmaceutical R&D investment in the U.S.A. represented less than 70% of that in Europe, but R&D investment in the U.S.A. has now overtaken that in Europe *(23)*. EFPIA believes that excessive interventions by European national governments to control pharmaceutical spending are denting confidence among multinational pharmaceutical companies to invest in the region.

It should be noted that innovation and productivity in all industrial sectors has been said to be lagging in Europe relative to the U.S. so price controls are really only one of many factors affecting the European research community.

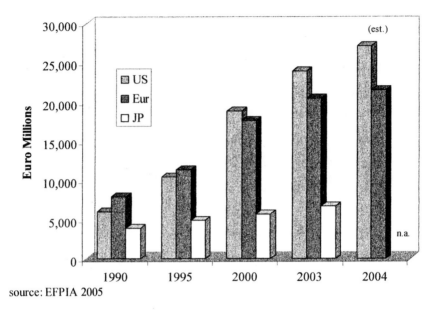

Figure 3. Pharmaceutical Expenditures for R&D in the U.S., Europe and Japan

U.S. R&D Dominance

As already mentioned, the U.S. has been the most effective at promoting conditions for new drug development in the last decade and when it comes to deciding where to invest in R&D, this has not gone unnoticed by companies operating on an international basis.

U.S. market conditions have enabled companies to maintain a long term approach to their R&D programs, whereas in Europe the frequent government interventions are seen as disrupting this approach. In 1977, the U.S. pharmaceutical industry invested around U.S. $1.3 billion in R&D, but in 2000 this figure had risen to $32 billion (2). As a result of this committed investment, the Pharmaceutical Research and Manufacturers of America (PhRMA) noted that eight of the current top ten worldwide prescription pharmaceutical products had their origins in U.S. R&D and that since 1990, the U.S. pharmaceutical industry has grown twice as fast as the overall national economy (1).

The U.S.A. has the leading global biotechnology sector, spends much more than foreign industries on R&D, and employs the bulk of scientists globally (Figure 4) (20).

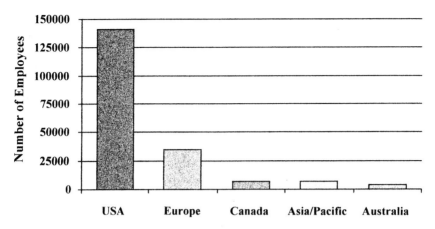

Figure 4. Employment in the Biotech Industry

In 2001, U.S. biotech companies spent U.S.$15.6 billion on R&D and accounted for over 70% of global biotech product revenues *(20, 24, 25)*. Moreover, the number of U.S. biotech companies is increasing. There are now close to 1,500 biotech companies employing 191,000 people. A 2003 survey of 850 U.S. biotech companies by the U.S. Department of Commerce revealed that the biotech-related technical workforce was growing at over 12% per year *(25)*.

The pharmaceutical industry interacts with many government bodies and financial communities that influence their activities. In the U.S., the federal government, through the National Institutes of Health and to a lesser extent through the National Science Foundation and Department of Defense, is the major source of investment in basic research directed to the benefit of public health. It dwarfs that of any similar institution in Europe or Japan. The National Institutes of Health (27) research budget is around U.S.$24 billion and the work of the NIH has significantly contributed to advances in the healthcare sciences *(1)*.

This research occurs at the institutes themselves, and through the funding of research activities at universities and small businesses. In 2002, federally financed chemistry and life science R&D at universities accounted for $13.5B, or 60% of total academic R&D spending in those specialties *[26]*. The rest of the financing was by academic institutions themselves, state and local governments and finally industry, which represented only 6.0% of funding. In addition, the government funding agencies support Small Business Innovative Research grants which, though monetarily small, have traditionally been seen as key investments for start-up companies. Current challenges to the policies which limit early investor participation in the companies while grants are in progress will most likely fall in favor of the investors and small companies given the government support for high tech high risk investing.

While the direct traceable impact of NIH funding on specific patentable discoveries and products is cloudy *(27)*, the enormous funding of basic research through this U.S. government agency is absolutely critical to the discovery of basic concepts in therapeutic targets and their validation, medicinal chemistry, screening technologies, computer modeling, genomics, and proteomics. The parallel benefit to this is that the funding maintains a high level of training at universities that then provide the key personnel for the innovation that occurs within industry itself.

The historical federal funding will not continue in the next few years at the same pace it has in the past due to changing priorities of the government in the face of national security concerns and a weakening economy. In 2006, the NIH budget increase from the previous year will be 0.7%, and less than 3% overall since 2004 which with inflation, will be a net decrease in funding. Coupled with a real dollar reduction in funding of the National Science Foundation (NSF) and a shift in priorities within the NIH funding to military and homeland security projects, this means reductions in funding to many academic institutions and other laboratories engaged in basic research that may ultimately benefit public health. It also means a reduction in education opportunities required to maintain the high quality talent pool that is attractive to first class pharmaceutical R&D. There is significant concern by the scientific community as a whole that there will be long term repercussions of the deprioritization of science within the United States and the eventual loss of status as the premier location for cutting edge science innovation in the world.

With many of the top universities in the world within its borders, the general availability of appropriately trained scientific personnel and state-of-the-art facilities has also contributed to the strength of the U.S. The output of PhDs has kept up with the population increase in the U.S., largely due to increases in foreign students supplementing the decreasing U.S. student participation. Based on a review by Freeman and Jin of the NSF Survey of Earned Doctorates, the percent of U.S. trained scientists and engineering at the top universities that are U.S. born has decreased from 77% of the graduates in 1973 to 59% in 2000 *(28)*. A more than doubling in the number of foreign born students and a bare 4% increase in the number of U.S. born students have occurred over the same period. The authors conclude that science has lost some of its ability to attract the brightest students because of opportunities elsewhere. What is also increasingly worrisome to the science community is that the foreign students particularly from Asia are finding more opportunities at home than ever before and the ability to retain their capabilities within the U.S. may be weakened in the future.

Despite the general stagnation of science jobs in the U.S., employment within the U.S. pharmaceutical sector is growing at around 4.5% (1, 2). But the flat-lining of the technology employment in the U.S. in general has made it difficult to speculate whether an employment shortage or surplus is on the horizon *(29)*.

Recently the National Academy of Science, the National Academy of Engineering, and the Institute of Medicine's joint Committee on Science, Engineering, and Public Policy evaluated the state of science activity in the United States. Their findings led them to be "deeply concerned that the scientific and technical building blocks of our economic leadership are eroding at a time when many other nations are gathering strength" *(30)*. They put forth four recommendations for changes in public policy (see Table III) and provided detailed examples of the current weaknesses and the potential courses of action to address their concerns. As will be seen throughout this chapter, the themes addressed in their analysis will be echoed by each region seeking to retain or grow their position in the R&D arena.

Cost containment is less of an issue in the U.S. because the government healthcare burden is less due to private insurance and a slightly younger population than in Europe and Japan. A study by the Massachusetts Institute of Technology Industrial Performance Center found that while the European governments of the UK, France, and Germany pay between 60 and 90% of their respective national drug bills, the U.S. government pays for only about 40% *(31)*. The authors concluded that the U.S. government had the least budgetary incentive to keep drug spending low. There is therefore a relatively high spending on pharmaceuticals by U.S. consumers, which makes the market of prime importance for pharmaceutical companies looking to launch new products *(31, 32)*. In the United States in 1999 for example, the average annual cost of healthcare was $4,271 which is almost double the cost in the EU countries as a whole *(18)*.

Table III. Report from the Committee on Science, Engineering and Public Policy

Recommendations
• Increase America's talent pool by vastly improving K-12 mathematics and science education.
• Sustain and strengthen the nation's traditional commitment to the long-term basic research that has the potential to be transformational to maintain the flow of new ideas that fuel the economy, provide security, and enhance the quality of life.
• Make the United States the most attractive setting in which to study, perform research, and commercialize technologic innovation so that we can develop, recruit, and retain the best and brightest students, scientists, and engineers from within the United States and throughout the world.
• Ensure that the United States is the premier place in the world to innovate, invest in downstream activities, and create high-paying jobs that are based on innovation by modernizing the patent system, realigning tax policies to encourage innovation, and ensuring affordable broadband access.

This is not to say that the impact of health costs on economic productivity has gone unnoticed. In the U.S., in 1998, public spending on healthcare represented only 5.7% of GDP, but private spending was 7.1% of GDP to result overall in the highest percent GDP spent on healthcare globally. The brunt of this cost has been felt directly by insurance companies, employers and the under-insured to date. The governmental unity of the United States has not been sufficient to make inroads into the growing dilemma even as the pressure on public insurance continues to grow. Recent publicity about cost differentials with Canada in the face of underinsured U.S. citizens with easy access to Canadian pharmacies has again rekindled the recurring discontent in the U.S. for having to shoulder the burden of R&D costs for the world.

In the end, any U.S. pressure on cost containment is not expected to achieve the level it has attained in Europe. There are many incentives why the pharmaceutical industry chooses to invest in the U.S. that override any concern about cost containment.

State Investment is Critical

Interestingly, state and local support of pharmaceutical R&D remains strong. The motives for this funding are less ambiguous and are clearly intended to make the state a more attractive place for pharma R&D to take place. The horizon is not far – states see competition mostly among themselves rather than with foreign countries, though some of the more robust states in pharma R&D recognize the outside competition more keenly.

In 2004, the Biotechnology Industry Organization commissioned a study by the Battelle Technology Partnership Practice and SSTI to conduct a survey of state government bioscience initiatives *(19)*. States were interested in the biosciences (life sciences) because of the higher than average pay for workers in that industry segment; the anticipated growth of employment in this area and the belief that bioscience research can benefit the public good. The majority of employment in the drug and pharmaceutical sector is located in seven states, each garnering over 5% of the total national employment in that industry sector (see Figure 5) *(24)*.

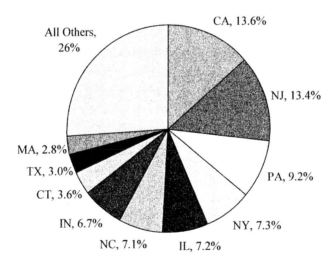

Figure 5. Employment in the U.S. Pharma Industry

There is definitely asymmetry in the distribution of the research environment even among the top R&D states. New Jersey,

Pennsylvania, and Illinois for example are home to large pharmaceutical companies that create their own R&D culture and environment. But California, Massachusetts and North Carolina are relative newcomers and have had to create the environments themselves, so-called "hotbeds," to be home to some of the ~1500 small and emerging life science companies in the U.S. that are largely research based organizations. It is interesting to note that the continual spin-off and start-up phenomenon seen in the hotbed communities is not shared by the large company environments.

Those states that are actively competing for pharma facilities, and in particular for pharma R&D facilities have recognized the need for providing capital, talent, facilities and a locally attractive business climate. This requires significant investment in university facilities and technology to attract world class basic research and medical researchers, creation of wet lab space in areas surrounding the universities, provision of a cultural environment that is attractive to the talent pool that has become highly mobile and selective in their living environment, providing facilities for non-profit research organizations to "seed" the area with an off-campus research culture and a ready pool of venture capitalists and investment funds in the locale. The significant investment also attracts federal dollars to the local area. There is a rough trend between federal (NIH) funding of bioscience and the number of biotech companies within a state. Table IV compares the NIH funding for institutions (19) with the number of biotech companies in the top states (33). These states constitute ~85% of the biotechs within the U.S. and attract 66% of the total NIH funding to institutions. The two most probably grow together rather than one driving the other. There are only a few states with substantive private sector employment in pharma R&D that do not have significant federal funding of research as well. Notably New Jersey, gets relatively little federal research funding even on a per capita basis, but replaces it with the largest employment in large pharmaceutical companies in the country. This large pharma community draws talent to the region and stabilizes the required environment for biotech research much like well funded universities do in other states.

Federal funding and private sector employment represent the outcomes of a variety of state efforts. All states with active pharma R&D activities participate in the venture funding of new enterprises directly or through private venture funds. All states invest in specialized bioscience facilities typically associated with universities, and in wet-lab incubator space for new entrepreneurs. Tax credits, grants and loans

Table IV. Relationship between NIH funding to institutions and biotech companies on a state basis

State	# biotechs	NIH funding to institutions ($M)	NIH funding $ per capita
CA	420	$ 2,904	$82
MA	193	$ 1,874	$292
N. Carolina	88	$ 781	$94
Maryland	84	$ 1,107	$203
NJ	77	$ 218	$25
NY	66	$ 1,714	$90
TX	64	$ 1,028	$47
GA	63	$ 311	$36
PA	63	$ 1,241	$101
WA	42	$ 674	$111
FL	33	$ 291	$17
CT	29	$ 391	$11

directly to new ventures are typically available though the magnitude of all of these financial enticements vary greatly among the states.

California – Home of Biotech

California is the quintessential biotech state with almost 24% of all U.S. jobs in biopharma oriented private sector research and testing, and home to two biotech hotbeds in the San Francisco Bay area and the San Diego area.

California's development sprang from their universities coupled with the entrepreneurial spirit and investor cash generated by the computer revolution of Silicon Valley 20 years in advance. The University of California claims that 1 in 3 California biotech firms was founded by UC scientists, and that 1 in 4 public biotech firms nationwide are located within 35 miles of a UC campus, albeit many of them are even closer to California's private universities like Stanford for equal reason. Clustered next to every major university is a private or non-profit research park, or planned research park, to allow rapid tech transfer and strong collaboration. Every key region has a trade association that stabilizes the network and can mobilize the community on key government issues of importance.

San Diego is a prime example of a region that has capitalized on its existing facilities, created a web of interacting companies, and generated explosive growth in the area in recent years (34). It is the third largest biotech hotbed in the United States, behind the San Francisco bay area and Boston. The growth of this community is characterized by an initial bioscience base of University of California at San Diego, Scripps Institute and the Salk Institute, among others, with significant federal funding. Influenced by the Northern California start-up and spin-off culture that came before it, and access to capital allowed the creation of in an every widening sphere community of life science activities. The region has become a mecca for large companies seeking technology value, including many foreign companies from Europe and Japan. In addition, several foreign venture funds including Novartis Bioventure Fund and JAFCO (Japan) have offices in the region.

The success has been accelerated and stabilized by the effective business – government relations that have supported the community. The regional business group for the San Diego and Southern California biotech community, BIOCOM, has 450 members who, aside from biotech companies, include law and investment firms, executive recruiters, non-profit groups and accounting, insurance and architectural firms that interact with the biotech companies themselves. BIOCOM offers a conduit for networking as well as being a representative of the community on local, state, and federal legislation that impacts the well being of the community.

Massachusetts – Another Innovation Driven Hotbed

Massachusetts' pharma R&D community shares the same self germinating history that does those in California. "Genetown" is anchored by some of the best universities and medical schools in the world and that see significant personnel exchanges with their California counterparts thus helping to transfer the same philosophy of pharma R&D activities. The proximity of Massachusetts to the home bases of many of the largest pharmaceutical companies has probably benefited the region as well. The strong relationship of large pharma with these small biotechs was underscored when Novartis chose to locate their newest major R&D facility to Boston in 2000, emphasizing the manufacturing, administrative centers or home territory. However, even this natural environment for innovative research still requires considerable support from state and local government to keep it competitive.

The state and quasi-public corporations provide capital through venture funds, direct seed-stage investments, and loans to small biotech businesses. The area is home to at least four research parks created by public-private development. Massachusetts' most significant universities are all private, unlike the mix in California, and thus the state participates in funding a smaller fraction of academic research institute development than elsewhere. As with all similar regions, there are several councils and organizations that assist in networking entrepreneurs and scientists with colleagues, with investors, legal firms, and assist in articulating political issues of importance to the community.

In 2002, the Boston Consulting Group and the Massachusetts Biotechnology Council issued a status report on the biotechnology industry in Massachusetts *(35)* and provided extensive comparisons with the California and North Carolina biotech arenas. The biotech industry has contributed about half of all new jobs in industry in the state and accounts for 27% of R&D activities and 18% of the state's venture capital investment. Given the enormous leverage of the mix of enterprises in the Boston area – universities such as Harvard, MIT and Tufts, world class medical schools, almost 300 biotech companies and the Novartis institute, the region has established one of the most solid biotech hotbeds in the world. However, its success has certainly spawned worthy competitors which are now of a concern to the area. Likewise, the report recognized several weakness in the Massachusetts region that could reduce the stability of their position in the long term. Four key challenges for the state, shown in Table V, were identified.

To meet these challenges, the report cited a number of familiar themes. Despite the unsurpassed educational institutes in the area, investment in science education was identified as an issue. Additionally, improving the financial climate of the area for biotechs was recognized. By participating in the development of physical spaces for innovation through zoning changes, planning, and funding, by providing tax credits and grants to start-ups and by appropriating state controlled money for venture funding, the region is expected to make the area more attractive to new and expanding enterprises. But the report also identified the need

Table V. Findings of MBC status report on Massachusetts

Key challenges for Massachusetts

- Respond to the growing competition from other regions
- Maintain the legacy of world-class research and innovation and become the best at converting research into commercial innovation
- Extend the industry from its base in research to activities further down the value chain such as development and manufacturing
- Leverage the resources and networks of the broader life-sciences economic cluster of which biotechnology is an integral part.

to get involved in the way research is done as well. Finding synergies between innovation fields of the "life science cluster" has the potential to accelerate innovative activity and find commercial value in a broader range of fields.

Commercial value was an important point raised by the report. Biotech enterprises in Massachusetts have been less successful than in California, for instance, in turning innovative ideas into commercial products. These concerns parallel the changes in definition of high valued biotech companies in the last 10 years. Originally companies with "platform technologies" in which innovative patents could theoretically generate a host of products were highly valued by investors. The lack of success by many companies in turning their property into any products has led to a rethink of company value. Today, goal oriented companies that can actually capitalize on their technology resulting in products and partnerships are the expectation. The report felt that the state could play a role in encouraging enterprises to reach maturity more rapidly by better focus and goal setting, and by creating an environment of business innovation. If Massachusetts companies as a whole cannot make the transition to the new paradigm, the overall value of the community will be less and will be more vulnerable to outside competition.

States that Purposefully Create Innovation Centers

The success of the first hotbeds, the loss of manufacturing jobs and the view that high tech jobs are less prone to globalization (i.e., transfer out of the country) has caused many states to attempt to create hotbeds in reverse. That is, by providing incentives to attract anchor businesses and institutes they hope to attract additional companies and investment, similar to the model of Research Triangle Park (RTP) in North Carolina. In that region, the research park was established early on in proximity to well respected medical centers, but growth was initially slow without native born key technologies. The establishment of anchor non-profit facilities helped to attract private enterprise and the area has expanded by the continual dissemination of scientists and entrepreneurs from one company to the next. However with the mergers and consolidation of several major pharmaceutical companies and contract research organizations with facilities in the region, and the competition from newer hotbeds, RTP may have a greater struggle in the future maintaining their position in the field.

The challenge is that the biotech hotbed ecosystem is complex and hard to re-create without an inherent demand, especially as the competition heats up for attracting world class researchers and investors. Three states which are trying to develop hotbeds are Florida, Michigan, and Arizona. Florida has managed to attract a large Scripps institute onto previously undeveloped land which will also allow for development of private enterprises as the demand arises. Florida has a considerably lower education profile than the California and Massachusetts hotbeds, but it is starting by growing science based bachelor degree programs at local colleges to provide adequate support staff for the institute.

Michigan, a traditionally industrial state suffered tremendously by the loss of transportation manufacturing in the last quarter of the 20th century. The state was more prepared for the loss of Pharmacia Upjohn when it ultimately was closed as part of the Pfizer acquisition in recent years. Michigan, which would have a difficult time developing a critical mass of a talented work force from scratch recognized that it had to create jobs for their highly trained, newly unemployed citizens, or quickly lose them. Again, the encouragement of partnerships with the state universities and colleges, the conversion of "abandoned" laboratory space to incubators, the creation of venture capital funds and direct grants to start-ups have all been employed to stabilize and grow the region. In 2003, 10 new start-ups were created by former Pfizer employees with the assistance of the state measures. It will take several years to discover whether these actions have led to the foundation of a new pharma R&D environment in the state or just slowed the flow of the talent pool to sunnier and coastal climes. Michigan is home base to several other technical industries that are being challenged from abroad include chemicals and transportation engineering. The state has chosen to consolidate efforts to shore up their status in technical innovation in all these areas; the success of such a plan is to be seen.

Phoenix Arizona is another region that is trying to bootstrap pharma R&D activities for the local economy. Home to Arizona State University, it has gotten an additional two state-run universities to collaborate on a new Medical School in the city area. The city has invested $46 million in the first of several new medical laboratories and has tapped into both state and private funding to expand the effort. They have two non-profit organizations and the National Institute of Diabetes and Digestive and Kidney Disease (NIH) as tenants. The hub has been able to attract several firms from outside the country. But Phoenix has had to compete with the more substantial hotbeds around the U.S. and presumably highly attractive offers were made to acquire these new members of their community. Whether they and similar sites can generate a center with enough momentum to go on its own and return the investment back to the community will have to be seen.

U.S. Private Investor Contribution to Stimulating R&D

The need for accessible private capital should not go unappreciated and is another area of strength for the United States, and certain states in particular. The U.S. has always embraced, perhaps more so than elsewhere, a culture of entrepreneurship, competition, and willingness to accept financial risk for financial gain. This culture is critical to the creation of a venture capital community that is interested in investing in high risk start-up ventures often required to fill the gap between academic research and large pharmaceutical companies. This group of investors supports small companies, often participating on corporate boards and mentoring the company until they develop sufficient intellectual property, unique skills and/or demonstrable therapeutic success to interest a wider investor community and established pharmaceutical companies. Without the ability to recoup their investments and make a substantial profit, financial support for start-ups would disappear along with the start-ups themselves. Even though VCs prefer to invest in local endeavors, it is the national stock market system that provides the next critical step. Public investors throughout the U.S. are ready and willing to invest in high tech stocks anywhere in the country, providing a wide base of investors. In addition, the NASDAQ exchange allowing trading of stocks that would have been shunned by the traditional stock markets, has been equally useful to support of the chain of financial opportunities which fosters pharmaceutical R&D in the U.S.

The financial maturation of biotech firms traditionally involved venture capital funding, additional private investments followed by IPOs and public funding through profitability. The lack of return on many of these ventures, the souring of the public market for high risk in general, and the increasing value of the proprietary technology to large pharmaceutical firms have led to a change in financial development of emerging enterprises. Over the last few years, venture funding has risen slowly but steadily while the number of IPOs and the value of those which have been completed have been erratic. Today the IPO market demands a more mature enterprise with more proven technology and that is closer to profitability than was tolerated five years ago. The gap has been filled by large pharmaceutical companies partnering at an earlier stage than ever before. This continues to make the initial private funding profitable enough for the process to continue onward. In 2004, about $31 billion was invested in U.S. biotech companies, or which about 13% was private funding and about 35% was through partnering *(36)*.

European Complexity

Although it is interesting to examine the state of the pharmaceutical industry in Europe, each country within the region is actually competing with others to attract companies and qualified staff (Figure 6) (23). Furthermore, within each of these countries there will be areas that compete with each other in similar way for resources and finance. Thus there is no single way of looking at the "European situation." Each European country would like to see more general investment by the pharmaceutical industry in the region as a whole, but would most likely prefer that it gets as much of this investment as possible (and a similar scenario would be the case within a country). It is impossible to examine every angle of European investment from a regional, national, and provincial level, but as long as generalizations are kept in perspective, 'top-level' analyses can be of value.

Examining Europe is also difficult due to the complex and changing political structure of the region. Many countries are members of the European Union (EU), but equally some are not.

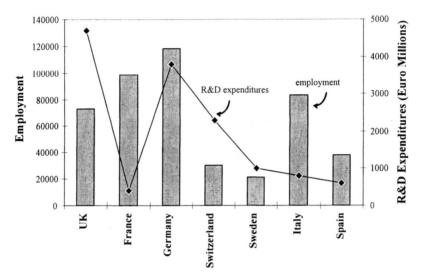

Figure 6. Pharmaceutical Employment and R&D Expenditure in Major European Industrial Centers (2003 – latest data)

Since being founded by Belgium, France, Germany, Italy, Luxembourg, and the Netherlands, various agreements have strengthened the concept of the EU. Enlargement has always been one of the important characteristics of the EU and it recently undertook its biggest enlargement ever in terms of scope and diversity and now consists of 25 Member States. The newest Member States add 75 million people to the 400 million already living in the EU *(37)*.

The expansion of the EU has generally been welcomed by the pharmaceutical industry as it widens their European market. Countries within the EU will at least have some operating conditions in common, due to what has been agreed at a political level, and so this reduces some of the variation seen across the region. Unfortunately, in the short term, EU-wide schemes to standardize conditions do not result in absolute harmonization and some differences remain. It is this lack of harmonization that makes operating in the EU problematic for companies. For every advantage that the EU offers as a potential single market peculiarities remain.

For example, despite the move toward more European integration and trade, the major European pharmaceutical markets of the UK, France, Germany, and Spain continue to have their own systems in place to set the reimbursement prices for pharmaceuticals *(32)*. The various national requirements that interface with EU-wide legislation, combined with varying cultural factors and expectations, continue to make it difficult for a company to derive commercial advantages from the EU market.

Financing of biotechnology is another area where the political fragmentation of the continent does not work in its favor. Each country has its own stock market with varying degrees of tolerance for the risk levels inherent in these enterprises. In addition, the stock markets do not capture as wide a range of investors as the U.S. stock markets which serve the whole country despite the specific location of the company. Thus there are inefficiencies and inequalities in the ability to obtain public investments. In addition, European investors, both private and public, are more risk averse than in the U.S. so that money overall is less available. Recently, BioEuropa published a commissioned report, contrasting the European biotech sector with that in the U.S. (38). Table VI shows key financial figures from the report for 2004, highlighting the disparity in the financial markets.

The disparity among the stock markets was also underscored in the BioEuropa report. While Germany and the UK constituted 26% and 23% of the total biotech enterprises in Europe in 2003, IPOs in 2003-04 were dominated by the London Stock Exchange and Alternative Investment Market (also in the UK). IPOs on the Frankfurt exchange were very meager in comparison to IPOs in EC countries with considerably fewer enterprises.

Table VI. Selective 2004 financial comparators for the European and U.S. biotech sectors

	Europe	U.S.
# of biotechs (2003 figures)	1976	1830
Venture capital	€940 million	€2850 million
IPOs	15	32
% total equity funds raised	16%	84%
% total debt provision	19%	81%

Aside from skittishness about investing, public opinion can have other subtle and not so subtle impacts on biotech investment and this is evident in Europe as it is elsewhere in the world. Europe as a whole has been cool to the "agrobiobusiness," specifically genetically modified organisms (GMOs). This is especially true in Germany. Research in GMOs in the U.S. is a respectable fraction of the overall biotechnology sector and allows for exchange of ideas, expertise, and tools with the pharmaceutically directed biotech efforts. The reluctance to engage in agrobiotech in Europe reduces the overall activities and the potential for synergies to foster innovation.

Mounting Cost Pressures

Mounting health care costs are certainly a major rallying point in the European community which shares a common philosophy of public healthcare. In May 2004, OECD Health Ministers held their first joint meeting to discuss the performance of the healthcare systems in their countries and to examine ways in which to improve them *(12, 13)*. Along with increasing life expectancy, spending on healthcare had also risen in OECD countries, with the average healthcare expenditure share of GDP rising from 5% in 1970, 8% in 2000 to almost 9% to date *(12, 13)*.

A three-year health project conducted by the OECD examining the performance of healthcare systems in different countries highlighted examples where the highest spending countries did not necessarily achieve the best healthcare outcome *(12)*.

The potential of expensive new technologies to improve healthcare outcomes and reduce expenses over time was recognized, but it was felt their benefits would need to be carefully assessed in the long term *(13)*.

Thus while one may expect less obsession with cost containment across the board in the future, longitudinal pharmaeconomic studies may become more important in acceptance and retention of new medicines. This may prove a double edged sword for the pharmaceutical companies. It is expected to support a wider use of chronic treatments to reduce the need for more expensive therapies, but may continue to delay introduction of newer medicines.

Boosting Pharmaceutical Innovation in Europe

Despite the difficulties in accounting for the variation across Europe there have been efforts to develop measures that could improve pharmaceutical innovation in the region as a whole. As the pharmaceutical industry is a major contributor to most European economies and is a major employer, its decline is in no one's interest. For example, in Europe, in 2001, the pharmaceutical industry employed over 560,000 people (with 88,200 of these working in R&D) and generated a trade surplus of €28,000 million. In fact, the pharmaceutical industry in Europe has been the only high technology sector to consistently show a growing positive trade balance *(39)*.

In late 2002, the G10 Medicines Group brought together top European industry and public health decision makers to consider ways of improving competitiveness in line with social and public health objectives. One of the important achievements of the group was the setting up a system of EU indicators that allowed comparisons to be made between the EU and its major competitors as a basis for establishing best practice within the EU and uptake in each Member State. The group also published a report on its findings for the European Commission, outlining proposals for concrete action to be taken, as Table VII summarizes *(40)*. The findings are interesting in that they parallel the industry's opinion that cost control is key to industry participation and that Europe needs to be more like the U.S. in its funding and regulation. This theme will be again seen when individual countries are discussed.

Early this year, the OECD published a report *(41)* examining the relationship between innovation policy and economic performance in six member countries of the OECD – Austria, Finland, Japan, the Netherlands, Sweden and the UK. While the report considered innovation in all technology sectors, it identified seven measures of innovation: macroeconomic performance, R&D spending, human resources for science and technology, scientific and innovative output, science-industry linkages, international linkages and technological entrepreneurship and industrial structure. While the strengths and weaknesses of the evaluated countries varied, the report concluded that they shared a need to profoundly change their innovation policies to be able to respond to new technological and economic developments.

Europe does face problems in staffing its pharmaceutical industry. Given that the most successful research organizations are based in the U.S., there is a general perception that opportunities for career progression may be better in the U.S. than in other countries *(42, 43)*. U.S. industry and academic institutes attract top scientific talent from Europe, Japan, India, China, and the former Soviet Union amongst others. The National Science Foundation has estimated that foreign students account for 40%

Table VII. Summary Recommendations by the G10 Medicines Group for Improving European Pharmaceutical Industry Competitiveness

Recommendations
• Stimulate R&D and innovation; • Strengthen EU medicines legislation to meet public health and industry needs; • Promote better access to innovative medicines; • Address the patchwork of national controls to create a more competitive EU environment. • Promote and stimulate centers of R & D excellence along similar lines to the U.S. national institutes of health to re-energize European research frameworks; • Improve medicines regulation encouraging more rapid decision making and efficient procedures • Improve data protection and maintain high intellectual property standards; • Enable full competition for medicines which are neither purchased nor reimbursed by the state. This could establish a genuine EU wide single market for non-reimbursed medicines • Provide mechanisms which give patients more access to high quality information on pharmaceuticals from industry as well as other sources.

of U.S. advanced degrees in biology and chemistry *(44)*. Even for those who eventually wish to return to their home countries, gaining a few years of work experience in the U.S. pharmaceutical industry can be a major advantage in terms of career opportunities on their return *(42, 43)*.

Whilst Europe has recognized the loss of their scientific talent to the U.S.A. for a number of years, its action to remedy the situation has been indecisive and slows *(43)*. In 2002, the European Council of Ministers called on EU Member States to devote 3% of their gross domestic product (GDP) to research *(44)*. It was estimated that if these measures were adhered to it would finance an additional 400,000 science jobs each year. Yet, while countries such as Sweden and Finland increased research expenditure, France actually reduced its research spending and recruitment of young scientists. In March 2004, over 2,000 leading French scientists and researchers resigned en masse and 70,000 scientists signed a petition to protest at government funding cuts *(45)*.

For example, even the following year, France's prestigious national science research centre, the Centre National de Recherche Scientifique (CNRS) is still owed half its research funding for 2002 *(45)*.

Other more unusual issues also stand in the way of European-wide schemes to improve the pharmaceutical industry's prospects at a regional level. For example, potential workers continue to have difficulties in moving between different EU countries *(43)*. It is not uncommon for a university in one EU Member State not to formally recognize qualifications obtained at a university in a different Member State. In the U.S.A., a degree from an accredited institution of higher education is generally accepted across the country, which enhances employment prospects *(44)*. Unless those seeking to improve the conditions in the EU also address these and related issues, there will always be a steady drift of high caliber individuals to the U.S.A. *(42)* or elsewhere.

The inconsistent performance of the sector has not been a help. If European biotech companies cannot emulate their U.S. counterparts in terms of graduate opportunities, there will be fewer incentives for students to consider biotech careers. Such a situation would further compound the problem of having a limited pool of high caliber staff entering the industry from within these regions. Although there are areas of promise within Europe, as a whole the region is still attempting to find the formula that will lead to a mature biotech sector.

Although EU-wide schemes to improve the conditions for the pharmaceutical industry are welcomed, many believe that national initiatives in the different Member States can have more of a dramatic impact. Critics believe that in the EU, the political set-up is not always conducive for effective legislation. With the current EU structure and its recent expansion from 15 Member States to 25, and more expansion anticipated, the process to develop legislation becomes lengthy and convoluted, as so many different views must be taken into account. They favor national legislation, which has a more defined focus, and follows a quicker route.

European authorities are trying to respond to the funding and productivity gap between their biotech sector and that of the U.S. The European Commission (EC) recently launched its Sixth Framework Program to encourage scientific research *(46)*. The EC Framework Program has a budget of €16,270 million (U.S.$16,400 million) aimed at supporting R&D in areas such as the life sciences, for projects with an eventual commercial objective. They hope that these initiatives will encourage small European companies to seek funding, as in the past they have been somewhat discouraged by the apparent bureaucracy in obtaining funding via such schemes.

The Environment for R&D in the UK

The UK has traditionally been a strong performer in terms of R&D for drug development and it is a position that the government wishes to maintain *(43)*. The Association of the British Pharmaceutical Industry (ABPI) reports that 15 out of the world's top 75 medicines were discovered and developed in Britain – more than any other country outside the U.S.A. *(43, 47)*. Moreover, the UK's biotech companies account for 43% of all biotechnology drugs in advanced clinical trials in Europe *(43, 48)*.

In many respects the UK has been the most proactive European country in trying to boost pharmaceutical and biotech innovation. Nearly 9% of current global pharmaceutical R&D is attributable to UK companies and accordingly this sector is viewed by the government as one of the most important contributors to the British economy *(2, 43)*. In fact whilst many countries in Europe have struggled to remain attractive to pharmaceutical companies, UK R&D expenditure has managed to sustain itself at a steady level. UK R&D expenditure as a proportion of estimated global R&D expenditure has remained relatively unchanged (within a range of 7-9%) since 1990 *(2)*.

The R&D environment in the UK has been strengthened through the work of the Pharmaceutical Industry Competitiveness Taskforce (PICTF) *(49)*. Set up in 2000, PICTF brought together representatives from industry and government to examine the steps that could be taken to make the UK more attractive for pharmaceutical R&D investment and to collect and publish annual competitiveness and performance indicators.

The British government has been keen to maintain the popularity of the UK as a research base for the pharmaceutical and biotech industries and is eager to increase the number of science students in higher education. In its 2002 report, PICTF revealed that whilst the proportion of young graduate scientists in the UK labor force was higher than in Germany it was below that in several other countries, including the

U.S.A., Japan, and France *(43, 49)*. Nevertheless, the report indicated that the pool of British graduate scientists, particularly those with biomedical qualifications, has grown steadily since the mid-1990s and that the UK labor market was generally perceived to be flexible and attractive *(43, 49)*.

Another advantage of the UK environment for R&D is in the way university-industry relations are organized. There is considerable industry funding of research projects in academic centers and a general atmosphere that promotes collaboration. Some observers believe that as companies are used to the fast pace of technological innovation they will be much better than governments in managing R&D productivity. Certain other countries have relied too much on public funding to try and drive innovation, but this is not seen as effective.

Another perceived benefit for the UK is the rigor of its financial markets. Investors have considerable experience of investing in the pharmaceutical and biotech sectors and are thus more realistic about the success of start-ups. In absolute terms this makes it difficult for a start-up company to find funding, but it suggests that those that do receive funding have a good chance of survival as their projects and strategies are considered viable by investors. For example, much of the coverage of the biotech sector in Europe has centered on the number of companies in different countries, but has focused less on their performance and long term productivity. Yet when examined in this way, the UK outperforms its European rivals *(2)*. At present, there are 18 profitable British bioscience companies with over 40 marketed products *(48)*. In addition, British biotechnology companies account for 43% of all biotechnology drugs in advanced clinical trials in Europe *(48)*. The biotech sector employs 25,000 people in the UK and this role as a major employer has motivated the government to create better conditions for its growth *(43, 48)*.

In 2003, the UK's BioIndustry Association (BIA) published a wide-ranging report by the Bioscience Innovation and Growth Team (BIGT), which focused on the national environment for biotech R&D *(50)*. The report was based on consultations with over 70 industry experts. The recommendations (see Table VIII) included programs to increase the scientific and managerial talent base available to the biotech sector. Further to these developments, a Bioscience Leadership Council was to be set up to oversee implementation of the recommendations and provide a forum for further initiatives.

Table VIII. BIGT Recommendations for the UK's Biotech Industry

Recommendations
• Build mutually advantageous collaboration between the UK's National Health Service and industry for patient benefit. • Create a public and regulatory environment supportive of innovation. • Ensure sufficient and appropriate funding is available • Build a strong bioprocessing sub-sector within UK bioscience • Develop, attract, and retain a high quality scientific and managerial talent base with appropriate technological expertise.

The UK should retain its position of strength within the global pharmaceutical industry if it can continue to attract the major companies *(2)*. A report commissioned by the ABPI showed that the UK's reputation amongst the major companies remained very positive. AstraZeneca, GlaxoSmithKline, and Pfizer accounted for more than 70% of total R&D expenditure in the UK and other companies with a significant presence in the UK included Merck Sharp & Dohme, Organon, Lilly, and Novartis *(51)*. In fact, employment in pharmaceutical R&D has increased ~50% since the early 1990s *(2, 47, 51)*.

Although the UK continues to perform strongly, there is serious concern that changes in the university system may limit the number of science students entering the pharmaceutical and biotech industry in the future and thus undo some of the recent efforts to boost R&D investment and improve staff training *(43)*. For a number of years, several universities have sought extra funding and have called upon the British government to raise the contributions that students make towards their education. While tuition is a staple of U.S. higher education, even state-run universities expect significant payment by students for their education, this is highly controversial in the UK.

The introduction of more student derived costs will certainly affect the student population at least in the short run as the economy re-adjusts to such a change. As the British government has committed itself to increasing the number of students in higher education and to improving the industrial R&D environment it is faced with a political dilemma. It is still too early to predict the impact of these educational reforms, but if these new measures do reduce the number of high quality science graduates in the UK, they would counteract the improvements being made by PICTF and BIGT (43).

Another recent problem in the UK has been the threatened closure of chemistry departments at certain universities which have experienced funding problems. The closure of the chemistry department at Exeter University, as part of efforts to reduce £4.5m annual budget deficit also represented 130 job losses *(52)*. The move attracted widespread criticism from parents and the pharmaceutical and chemical industries, who have brought pressure upon the government to address the underlying funding problems *(53)*. Many observers have criticized the short term approach in dealing with the problem. The Royal Society of Chemistry commented that although chemistry was a more expensive subject to teach, in the long-term it resulted in greater economic value than other subjects. The ABPI has warned that a national strategy for key academic subjects, such as chemistry, must replace the current situation where local university finance, and funding councils that do not acknowledge industry's requirements *(54)*. Otherwise they believe that there will be a drop in the number of qualified individuals available to the industry.

The Environment for R&D in France

Historically, France has been one of the major forces in the European pharmaceutical industry (Figure 7) *(23, 55)*. According to the French pharmaceutical industry body, LEEM (Les entreprises du médicament), since 1995, France has been the leading drug producing nation within the European Union. LEEM members account for 98.7% of pharmaceutical R&D activities in France *(55)*. On a national basis, France remains the third largest producer of pharmaceuticals worldwide and it has been attempting to maintain its strong position despite an unfavorable economic climate.

A recent survey by the Organisation for Economic Co-operation and Development (OECD) revealed that pharmaceuticals accounted for more than 10% of total health spending in most countries, but that in France they accounted for over 20% of healthcare spending *(56)* despite the fact that it has one of the strictest pricing systems in Europe *(32)*. A combination of low prices and a high level of government-reimbursement have given patients and physicians little incentive to cut consumption.

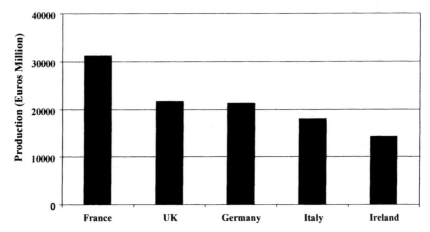

Figure 7. Pharmaceutical production in Europe (2003 – latest data)

With government policies exerting downward pressure on profitability and a notable increase in company mergers, France's industry is now at a crossroads. The change in the French pharmaceutical landscape has been dramatic. In the 1950s there were around 1,000 companies involved in the sector and this dropped to around 300 in 2002 *(55)*. On the one hand there are positive signs for France's R&D future. For example, LEEM estimates the numbers of pharmaceutical staff have tripled in 20 years with 1,000 new posts being created by the industry every year. At present, the industry employs around 100,000 personnel with about 18% of these being involved in R&D functions. The number of personnel involved in R&D has increased from 14% in 2001, indicating favorable conditions for pharmaceutical research in France.

In the past year the French Government has launched several initiatives to sustain innovation, including reforming the R&D tax credit scheme and creating a new fiscal status for emerging innovative companies. All are designed to improve France's competitive position in the international healthcare industries and make the most of available opportunities. France has been examining initiatives launched by the UK and Spain to improve their R&D positions. The French government has also been following the progress of the EU-focused G10 Medicines Group, which seeks to improve R&D competitiveness in line with social and public health objectives *(39, 40)*.

In January 2004, the French government commissioned a fact-finding report entitled PharmaFrance 2004 detailing measures that could improve France's R&D position. To compile the report a delegation visited several countries representing major R&D centers in order to hear the views of different companies and organizations involved in the pharmaceutical and biotech sectors. The report summary (Table IX) principally recognized the need for government – industry collaboration in sector development. These recommendations were incorporated into the PharmaFrance 2004 report *(57).*

Table IX. PharmaFrance Recommendations for the French biopharmaceutical sector

PharmaFrance Recommendations

- Create an annual fund, of €100 million upwards, to fund joint public-private projects
- Negotiate with the pharmaceutical industry to find common ground over R&D policy and pricing issues
- Create a task force involving Ministers and industry representatives to meet every 6 months and examine measures to attract R&D-based companies to France
- Promote state and industry dialogue to identify and overcome administrative and organizational obstacles to effective cooperation.
- Promote state and industry collaboration to enhance France's attractiveness as an environment for clinical research.

Perhaps one of the less welcome forays of the French government into the pharmaceutical sector was over the fate of Aventis in early 2004. Although Aventis has been described as a Franco-German company (as a result of the 1999 merger between Germany's Hoechst Marion Roussel and France's Rhône-Poulenc Rorer), in the context of its future, the company's French origins appear to have dominated.

Aventis, in need of a merger, had rejected an offer by its French compatriots Sanofi-Synthélabo and was reportedly considering an alternative approach by Swiss firm Novartis. However, the French government very publicly opposed any merger with a foreign partner and apparently placed pressure on the two French parties to come to an arrangement, thus paving the way for the creation of Sanofi-Aventis.

Investors appeared less convinced than the French government about the long-term prospects of the merger and were unhappy that the outcome of the merger was dictated by politics more than finance. However, the PharmaFrance 2004 report on the state of the French pharmaceutical industry suggested that Sanofi-Aventis would represent a resurgence of France's pharmaceutical sector as it takes up its position as the third largest pharmaceutical company in the world *(57)*.

The Environment for R&D in Germany

A prime example of the European pharmaceutical industry's concerns over the environment for innovation is illustrated by its experiences in Germany. The decline in the country's pharmaceutical industrial position has been dramatic and suggests an uncertain future for its reputation as a centre of R&D excellence.

Figures suggest that Germans pay more for healthcare than any other country apart from the U.S.A. and Switzerland *(58)* and therefore the public are opposed to any further increase in what they must pay for medical care. The pharmaceutical industry believes that the government's preoccupation with cost containment has weakened the underlying environment for R&D in Germany and has underestimated the industry's contribution to the German economy.

In its own words, the German Association of Research-based Pharmaceutical Companies, the VFA, has described the cost containment policies of the government as putting Germany among the "also-rans" in terms of competitiveness *(17)*. For example, it said that in 1990, Germany represented around 9% of global pharmaceutical production, but by 2000 this had dropped to 6%. In 1997, Germany was the European leader in terms of pharmaceutical R&D spend, it is now behind both France and the UK resulting in a significant loss of current and future revenue *(2, 17, 23)*. While a significant amount of the R&D now occurs in small and emerging biotech companies, Germany has not kept up in this arena (Figure 8).

In 2002 the VFA found that almost half of its member companies intended to decrease their R&D expenses in Germany during 2003, with another quarter planning on freezing R&D spending *(17)*. For example, Pfizer estimated that they could lose up to U.S.$164 million in annual

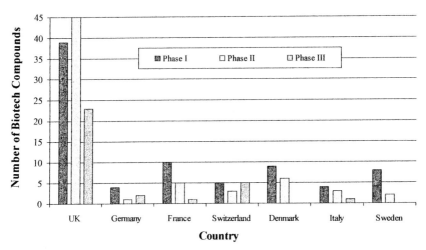

Figure 8. Clinical Development Profile for Biotech Compounds in Europe

revenues because of the new government reforms and decided to relocate certain staff to its UK operations whilst also instituting a hiring freeze in Germany *(59)*. Similarly, Merck & Co. decided not to proceed with the building of a new research complex in Munich.

From 1993 to 1995, the VFA member companies lost approximately 6,500 jobs in the sector. The German pharmaceutical industry remains a major employer as indicated by figures from the European Federation of Pharmaceutical Industry Associations (EFPIA) showing that it represents around 20% of the European pharmaceutical workforce *(39)*. Interestingly, the main beneficiaries of the fragility in the German pharmaceutical sector will be other European countries, particularly as the governments in the UK, France, and Spain are all actively seeking to attract investment.

The German government did put forward a plan to improve industrial conditions, entitled "Agenda 2010" in which it promised to remove barriers to investment, cut taxes, and create greater flexibility in the labor market. According to the document, conditions in Germany are not as bleak as often described and are improving. For example, it is noted that Germany is responsible for 6% of global pharmaceutical production and that between 1996 and 2002 production increased steadily *(60)*.

The Environment for R&D in Spain

Spain is an example of a European country which has not been a strong player in the past, but seeks to capture some part of the industries that are looking over the borders from their traditional locales. Spain currently occupies the fifth place in Europe and the seventh in the world in volume of pharmaceutical sales *(39, 61)*. Between 2001 and 2002, the Spanish pharmaceutical market grew by over 10% and is predicted to maintain a double-digit growth rate over the next five years *(61)*.

As well as a buoyant pharmaceutical market, Spain is a growing centre for pharmaceutical R&D. The pharmaceutical sector is widely considered to be the most innovative industry in Spain.

Figures from the European Federation of Pharmaceutical Industries (EFPIA) and Farmaindustria, the National Association of the Pharmaceutical Industry in Spain show that there has been a rapid rise in R&D investment in the country since the early 1990s *(39, 62)*. In 1990, pharmaceutical R&D investment was below €150 million but by 1999, this had increased to over €350 million *(39, 62)*. By the end of 2004, R&D investment was predicted to be around €485 million *(62)*.

The growing importance of pharmaceutical R&D to the Spanish technology sector and thus the national economy is illustrated by comparisons with R&D for other industries *(62)*. Between 1999 and 2002, there was a 31.5% increase in R&D activity for the Spanish pharmaceutical industry. However, over the same period, R&D activity for the aerospace industry decreased by 2.4%, there was a nearly 11% decrease in R&D activity for the automotive industry and R&D activity for radio, television and communications dropped by nearly 31%. Because of the general economic downturn that has troubled other industries in Spain, the pharmaceutical industry's share of the total national R&D activity in the manufacturing sector rose from 13.7% in 1999 to 17.9% in 2002.

The pharmaceutical industry in Spain is a major employer in the technology sector. At present there are around 270 pharmaceutical companies with production activity in Spain and approximately 375 laboratories *(63)*. These companies employ close to 39,000 people, which represent about 7% of the total European pharmaceutical industry workforce *(39)*. In 2001, approximately 9% of staff was employed in R&D functions, representing just over 12% of total research employees of industrial companies *(62)*.

According to the Spanish Economics Ministry, almost 90% of pharmaceutical companies are located in the Madrid and Catalonia autonomous communities *(64)*. Most of the companies surveyed by the Ministry were considered as "small to medium" enterprises, with about 35% of them employing from 100 to 250 people and only 2% having more than 1,000 employees.

Since 1985, the country has seen increasing investment from the large multinational pharmaceutical companies. The influx of investment from multinational companies was partly due to Spain having joined the European Union and having implemented full patent protection on branded drugs. It is estimated that foreign companies now represent 75% of pharmaceutical producers in Spain *(64)*. In 2002, eight multinational companies featured amongst the top ten leading companies in Spain and accounted for over 50% of the national market *(61)*. Only two Spanish companies, Almirall and Esteve, featured in this list.

Another reason why the pharmaceutical industry has invested in the Spanish market has been the availability of experienced and talented staff at competitive wages *(63)*. Spain has the second highest percentage of higher education students within the EU. It has been estimated that the labor cost of scientific personnel is approximately 35% lower in Spain than in the U.S.A.

As with other countries in Europe, the Spanish government wishes to reduce, or contain its own expenditures on healthcare, while attracting the pharmaceutical industry, particularly R&D enterprises, that can lead to an increase in personal income, national GDP, and exports. Spain has taken the approach of setting goals tied directly into national economic welfare and then rewarding companies that contribute to these goals, summarized in Table X. Companies that participated in the scheme were assigned a grading. The grading had important implications as it potentially translated into financial support through direct subsidies or zero interest loans.

An analysis of the success of the program was issued in 2002 *(65)*. To a degree the Profarma program was successful as it involved 56 pharmaceutical companies, whose product sales represented 85% of the Spanish market. Although financial support was possible because of the grading system used, it has been argued that the main benefit from the scheme was in terms of enhancing company's favorable status with the government, which was helpful when products were undergoing regulatory and pricing approval.

Table X. Profarma's goals to support innovation in the Spanish pharmaceutical sector.

Goals to support innovation
• R&D expenditure as a percentage of ethical drug sales should reach 8%
• R&D investment and expenditure should rise to €312 million
• There should be at least 3,000 employees in R&D roles
• The pharmaceutical trade balance should be improved

This is probably one of the more direct linkages of pharma investment in an area to enhanced localized profitability. While the sustainability of such initiatives is hard to foresee, holding a piece of the European R&D activities in the short term provides an opportunity to create a self sustaining R&D community around it. At the very least, it will buffer the country during a period of radical economic change.

Japan

Japan is certainly not a European country, but its position as a developed country and the presence of a mature pharmaceutical industry make it more similar to the U.S. and Europe than to the rapidly emerging biopharma communities of the rest of Asia. Japan's industry spends almost 10% of its revenue on R&D. The track record of Japanese companies in producing pharmaceutical products that have had a global impact is evidence of their position ahead of the Asian newcomers *(66)*. A number of products marketed by U.S. and European based companies have their origins in Japanese R&D include Bristol-Myers Squibb's Pravachol (pravastatin) for high cholesterol, TAP's anti-ulcer drug Prevacid (lansoprazole), and Daiichi's antibiotic Levaquin (levofloxacin). More recent examples include AstraZeneca's lipid-lowering drug Crestor (rosuvastatin calcium), which originated at Shionogi; Bristol-Myers Squibb's Abilify (aripiprazole) for schizophrenia, the result of a collaboration with Otsuka; and Eisai's Alzheimer drug Aricept (donepezil hydrochloride), co-marketed with Pfizer *(66)*.

Like other industrialized countries, Japan's rising expenditure on healthcare has implications for its position as a major R&D base. The trends in healthcare spending have led to the industry coming under pressure regarding its prices and this shows no sign of abating.

As with the pharmaceutical industry in Europe, the Japanese pharmaceutical industry feels that it is being unfairly targeted by such approaches and has warned that the country's position as an R&D center is being damaged *(67)*. There has been a growing trend for Japanese companies to establish research centers and manufacturing operations abroad in order to strengthen their position in foreign markets. In 2004, Daiichi announced that it would use New Jersey as its global clinical drug development operations, indicating an R&D shift away from Japan. Its merger with Sankyo is in line with its plans to globalize, particularly in order to capitalize from the favorable conditions in the U.S. market, and so a further flow of R&D investment out of Japan is to be expected. Other mergers taking place within the Japanese pharmaceutical sector could accelerate this trend.

Another issue for pharmaceutical companies has been the Japanese regulatory review timelines, which until recently, had become increasingly protracted over the last several years. Indeed, some pharmaceutical companies that have marketed their products in Japan have experienced timelines as long as 3 years for obtaining marketing authorization *(67)*. The fragility of the domestic research environment is also illustrated by the fact that there has been a steady decline in the number of clinical trial applications made in Japan not only from foreign applicants, but also from local companies because of increasing costs and lengthy review timelines.

Despite Japan's productivity on the pharmaceutical front, its biotech sector has lagged behind that of the U.S. and Europe. Although there is a high level of academic research that falls into the biotech field, it has not been geared towards commercialization.

One of the problems has been that the concept of the start-up along U.S. or European lines has attracted few Japanese researchers due to a lack of clarity in the regulations. However, there has been a recent change in the law that allows faculty members at national universities to serve simultaneously as corporate executives in start-ups *(69)*.

Funding such ventures is also a problem. In 2003 Japan Bio-Venture Development Association's (JBDA) launched a collaboration with Global Tech Investment in order to increase funding for start-ups *(70)*. This initiative aims to connect technology-based researchers in universities with interested parties in the business community. The

JBDA is in charge of identifying potential technologies in the university environment whilst the GTI concentrates on developing funding through its business specialists.

The Japanese government is now attempting to respond to reverse the erosion of the R&D environment, whilst sticking to its commitments to improving healthcare provision for its population. Recently, the government and the Ministry of Health, Labor and Welfare (MHLW) outlined their ideas to revitalize Japanese R&D through their "Vision of the Pharmaceutical Industry" and "Outline for Biotechnology Strategy" initiatives *(68)*. There are also initiatives underway to encourage people to take up careers in industry research. A target of 2010 has been set for the transformation of the Japanese R&D environment and although progress is being made, Japanese companies face intense competition from foreign companies seeking to increase their market share in Japan.

Emerging Asia

Many Asian regions are rapidly expanding to compete as centers of innovation. No significant discussion is possible here as each would take a whole new chapter in itself. A cursory review of some of the locales and their approaches is provided.

It is no coincidence that many emerging countries are developing their pharmaceutical sectors in order to attract investment from multinational companies. Many of the major established world economies view a trend with unease.

The region is a contrast in the very big (India and China) to the very small (for example South Korea, Taiwan and Singapore) but share certain similarities. They are typically characterized by large scientific talent pools or large *potential* talent pools if they can attract back many of their foreign-trained nationals. For example, Taiwanese nationals returning from the U.S.A. started Taiwan's Hsinchu Science Park, which features a number of young biotech companies [SIPA]. In 1986, biotech firms in Hsinchu employed 254 people, but this had risen to 823 by 2003 *(43, 71)*.

The countries typically have a wide variety of sophistication in the available facilities some of it being extremely limited by modern standards, and a limited domestic financial base for innovative ventures. Foreign investors are a significant source of capital. However, as the examples below show, the countries differ in the role of government and the organization of the overall strategy of development.

India Invests for the Future

The Indian pharmaceutical industry has openly expressed a desire to globalize. Its low-cost production facilities and large science base have made it an attractive prospect to other countries seeking cheaper pharmaceuticals. Up until recently, India was known as a major innovator in process chemistry rather than development of new molecules and therapies. This was driven by their successful business model of selling low cost generics into the U.S. and European markets, as well as providing pharmaceuticals to their own population and to many other developing countries.

India has up until recently depended upon the India Patent Act of 1972 and the Patent Rules of 1974 *(72)* for determining the scope of intellectual property rights. This protected drug manufacturing processes, but not necessarily the molecules themselves, thus giving them free reign to optimize processes. India has adopted international patent standards this year which will probably have an impact on their relationship with the developed countries, but lack of true enforcement early on might continue to allow a flow of "illegal" products to poorer countries that are less concerned about protecting their own rights.

Indian pharmaceutical manufacturers have certainly been innovative in their business approaches. Rather than accepting second class world status in the face of the large U.S. and European manufacturers, they have chosen to break out of India and take up positions of innovation in both regions themselves.

The Indian companies that appear best positioned to achieve international success are Ranbaxy, Dr Reddy's Laboratories, Sun Pharma, and Cipla. In particular, Ranbaxy has entered a period of rapid expansion and is ranked 70 in terms of global pharmaceutical companies and in 10th position in terms of global generics. It markets 28 global brands in over 70 countries and in 2002 achieved worldwide sales of U.S.$969 million. Ranbaxy has set itself a target of 2012 to achieve 40% of its overall revenue from proprietary prescription products, whilst also becoming a top 5 global generics company *(32, 72)*.

In 2003, the Indian pharmaceutical industry spent U.S.$147 million on R&D, which represents a doubling of its investment since 1999 *(73)*. Indian companies still have several challenges ahead of them but they appear set to be an exciting force in global drug development.

As discussed before, biopharma R&D and innovation requires talent, capital, and facilities, all in the critical mass and mix to become self-stimulating. India educates, and sends overseas for education, more scientifically trained people than they can gainfully employ, which depresses wages and allows India to compete dollar for dollar with the developed countries. The common shared English language has eased their entrance to the global community as well. Like other regions, they have government-funded support agencies and have established research parks, such as in Hyderabad, and technology focused institutes at universities. But India's development strategy includes a significant influx of foreign capital to grow. They have largely taken the route of being a contract service provider rather than an intellectual property generator and have thus expanded through the support of R&D companies in the U.S. and Europe rather than direct investors.

But opportunity can cut two ways. The traditional pharmaceutical companies recognize the significant market that India, with a population of 1.08 billion, is rapidly growing amid middle class expectations of health. Thus establishing market bases and clinical study centers is on the rise. In addition, the theoretically controlled intellectual property rights and the low wages of a highly skilled workforce has encouraged the establishment of U.S. and European based R&D centers in India as well, including medicinal chemistry, screening and toxicology laboratories. A number of U.S. companies, large pharmaceutical companies, contract research organizations and emerging biotech companies, have set up "back rooms" in India. In these cases, the business offices and top technical personnel remain located in the U.S. and do business directly with U.S. companies. The laboratory work is sent to Indian facilities, either operated by the U.S. company, or as contracted work for hire. Such a system was almost inevitable given the large number of Indian nationals carrying out pharmaceutical research at U.S. universities, institutes, and industries over the last 20 years and the quality of several of India's universities.

Singapore

Singapore has a well developed government strategy for positioning themselves as a world player in biomedical research. It has focused on education, facilities and investment as well as creating the attractive work environment that can compete for international talent *(74, 75)*.

Singapore's Biopolis is a completely boot-strapped, state-of-the-art mega facility for biomedical research. Initiated in 2000, the 2 million sq ft R&D complex when finished will eventually house about 2000 researchers. The state has created a variety of institutes focused on key research areas such as the Center for Molecular Medicine and the Institute of Molecular and Cellular Biology.

It is already home to the Novartis Institute for Tropical Diseases and several other enterprises.

Singapore has less of an advantage than many of the other Asian nations when it comes to available personnel. It has placed a great deal of emphasis on education and staffing for its emerging life science sector. It has set up the Agency for Science, Technology and Research (A*STAR) to integrate public research with developing industry clusters, and to ensure that they have a pool of suitably qualified staff to draw from. According to the Singapore Economic Development Board (SEDB), tertiary institutions produce around 35,000 graduates every year with qualifications suitable to the technology sector. In addition, the country has taken several steps to bring in world class talent from elsewhere. The Biomedical Research Council (BMRC) established an International Advisory Board with highly respected international scientists. North America for Contact Singapore is a government agency that recruits scientists from around the world and to date the success of recruitment and development of international alliances has been truly impressive including S*BIO, a partnership between SEDB and Chiron Corporation, the John Hopkins Singapore biomedical division, and the Singapore Onco Genome Laboratory at Biopolis a joint venture between the Max Planck Society and A*STAR, the last two of which have brought high level scientists from their partnering facilities to head the Singapore sites. In 2003, nearly 6,000 people were employed in Singapore's pharmaceutical sector, which represented an increase of 12% over 2002 *(43)*.

Finally, the BMRC has strategically created the facilities and institutes, and recruited personnel to vertically integrate the biomedical research activities from basic discovery work, through preclinical studies and clinical research in ground breaking areas. This will provide the greatest opportunity to obtain optimum monetary return on the government investment.

China

International optimism surrounding the growth of China's economy has attracted interest in its biopharmaceutical sector. Recent years have seen the establishment of a large number of domestic pharmaceutical companies as well as an influx of multinationals. According to BCG Analysis 2002, China is expected to become the fifth largest pharmaceutical market in the world by 2010. Companies such as Roche and Eli Lilly have set up R&D centers in the country and others have outsourced their R&D work to Chinese enterprises.

Talent is a key factor for the rapid development of China as a biotech force. The number of science graduates is increasing much faster than in the U.S. and Europe and it can entice back many foreign educated citizens. The extremely low salaries allow the enterprises to offer services back to the U.S. at costs that would be impossible to compete with in the developed world.

One of the worries of foreign investors is that although the Chinese economy has been growing at a rapid pace, unless its growth is carefully managed it could run out of control. This situation has occurred in other emerging markets but there are signs that the Chinese economy is more robust. Optimists point to the fact that unlike other emerging national economies that have run into problems, China has a current-account surplus and little foreign debt *(76)*. In particular, China's admission into the World Trade Organization has been seen as a major boost to future economic performance *(77)*.

The environment for intellectual property protection is still a concern in China. Pharmaceutical companies have been alarmed by the 2004 legal decision to overturn the Chinese patent for Pfizer's Viagra (sildenafil citrate). This was followed by a case where GlaxoSmithKline stepped back from defending its patent in China for rosiglitazone, the active ingredient of its anti-diabetes drug Avandia.

Ironically, there are some observers who believe that these problems will act as a driver for the establishment of a better Chinese system for intellectual property *(78)*. The United States Patent and Trademark Office (USPTO) is working with the Chinese government to help the government develop the appropriate guidelines and legislation in all areas of pharmaceuticals. Furthermore, in February 2005, the USPTO held a series of seminars in the U.S.A. to introduce the Chinese criminal justice system to intellectual property owners and other interested parties *(79)*. It was hoped that this might help U.S. companies better understand how Chinese legislation differs from systems they were more familiar with and how they could use the Chinese system to their advantage. Thus multinationals are likely to proceed with caution as they expand further in China.

No country has more direct involvement by government in the pharmaceutical enterprise than China. Most Chinese pharmaceutical companies are state owned and at present, 60 leading companies generate 70% of the entire industry's profits *(71,80)*. Due to the poor overall performance of its pharmaceutical sector, the Chinese government has been attempting to streamline the 6,000 companies that operate in the country and transform them into more productive enterprises.

In 2000 China was second only to the U.S. in terms of pharmaceutical production capacity, but this only represented 5.7% of global output *(80)*. Since then, production capacity has grown considerably. In 2003, annual production amounted to $47 billion, after several years of double digit increases.

Strategic leadership is one of the key weaknesses in the present day Chinese biotechnology sector. While the centralized government system may be very effective at streamlining and modernizing the manufacturing industry, its strategy and implementation for nurturing biotechs has been less efficient. There are several bodies within the Chinese government that see themselves as overseers of the biotech community and have separately developed strategies for growth. Last year a leadership committee for biotech development was established to harmonize the different ministries but it is unclear whether this will resolve the leadership issues or not *(81)*. In other countries, private business councils could fill the leadership void but this is not as facile in China.

China has a very large government investment in biotech through direct ownership or state-owned venture funds, but limited private capital. The Chinese government expected to invest about $1.45 billion in the biotechnology sector in 2001-05 *(82)*. While there are over 1000 biotech research facilities at present (numbers vary considerably), only about 30% of them are privately owned. In June of 2005, the first international venture fund was formed, directed toward investment in China. BioVeda China Fund, expected to raise up to $30 million initially included institutional investors and the World Bank Group's private sector arm.

Like other countries though, clusters of biotech enterprises have developed, notably around universities and institutes in Shanghai, Beijing, and Shenzhen. The government has also established a number of focuses research institutes. China has a relatively fluid interaction between small commercial enterprises, institutes, and universities, with about 5,000 pharmaceutical R&D facilities in the country. Those seeking inexpensive research personnel and facilities in China today can find universities labs competing with small entrepreneurs for a piece of the contract research opportunity. Chinese nationals have integrated very strongly into the U.S. pharma R&D community in all subfields and at all levels. A number of them have started enterprises with a foot on both continents – interacting with clients, investors, and key scientists in the U.S. while carrying out the laboratory work in China *(83)*.

China remains a strong agriculturally based country where productivity has far to go. Thus the biotechnology sector includes a large effort in agrobio research and development as well which may lead to China have strengths where synergies are identified.

In particular, the scale of R&D investment for producing novel drugs is currently beyond the reach of most Chinese companies. The top multinational companies spend around 15% of their revenues on R&D but the corresponding figure for Chinese companies is in the low single figure digits *(79)*. Furthermore, Chinese companies will need to change their commercial approach to drug development. To date, managers have been focused on the short term profits that can be generated by generics rather than the longer term potential profits arising from innovative research.

Stem Cell Research – an Example of Global Competition

One cannot talk about drivers for innovation without discussing the political framework of stem cell research. Because this report is being written in 2005, the topic falls under California, but several years ago this would not have been the case. Governments have often tried, for good and bad reasons, to direct what research is carried out by targeting funding to specific areas. But in a world with many parties vying for a key role in the innovation of pharmaceutical R&D, the research community will not stop at borders and will find places that are willing to support innovative work regardless.

Research in stem cells, that is, unspecialized cells that can differentiate into a variety of cell types, offers the promise of replacing damaged tissue to restore function. The potential opportunity to "cure" those permanently debilitated by spinal chord damage, stroke, Type 1 diabetes, kidney and liver disease, and Alzheimer's Disease, to name a few, is enormous. Current knowledge leaves much work to be done, although no one would dispute that this is a worthy goal. Embryonic stem cells, are thought to have much more potential for full differentiation than adult stem cells. However, the use of embryonic stem cells provides moral, ethical, and legal dilemmas that vary tremendously, most strongly from a religious perspective. This has led to a country-specific restriction of government funding for basic research as well as banning all research on human embryonic stem cells in some locations.

In 2001, the European commission held a discussion to review the scientific and ethical issues surrounding embryonic stem cells and found significant differences in how countries viewed the issues. Much to the dismay of some countries within the EU, the commission recommended the use of European Commission funds for basic research, albeit with careful monitoring *(84)*. Germany, Italy, and Spain for example ban such stem cell research within their borders, while the UK was less ethically compromised. Various groups in the EU mobilized to ban such research throughout the member countries.

At about the same time, the Bush Administration in the United States elected to ban the use of federal research funds for exploration of any embryonic stem cells except those already available at the time of the ban. The general feeling of the scientific community was that these cell lines were not of clinical value, severely limiting their ability to caring out high level research and effectively shutting down progressive stem cell research in the U.S. Despite that fact that there are many private funders of research, most of that money goes to development or applied research in which the return on investment is clearer.

The UK and several Asian countries such as South Korea recognized the opportunity to capitalize on the shrinking funding for this research as a means to strengthen their own research communities. In 2002, the UK Medicines Research Council launched a stem cell initiative with £40 million in national funds and additional funding by patient groups including the establishment of the first stem cell bank, forums, grants, student scholarships, and faculty fellowships. The Institute of Stem Cell Research headed by a key U.S. researcher Rogen Pedersen, attracted to the favorable research climate, highlighted the disparity between the U.S. and the UK.

"The strategic grants will ensure that the UK is at the forefront of the international research community working on stem cells, and is in a position to lead on the considerable health and economic implications the field promises for the future," stated the MRC *(85)*.

However, U.S. states recognized that they could use the lack of federal funding to differentiate themselves from their competitors as well and began increasing funding themselves. Many states earmarked special funds for such research.

Then in 2004, California passed the Stem Cell Initiative in 2004 which provided $3 billion dollars in funding over 10 years. An economic analysis of the initiative *(86)* concluded that it would provide $6.4 to 12.6 billion in state revenues and healthcare savings costs during the payback period. The returns were estimated as: (a) direct income and sales tax revenue of at least $240 million from spending on research and facilities; (b) additional tax income of up to $4.4 billion if a 5% increase in private investment occurred because of the initiative; (c) direct health care cost savings to the state of at least $3.3 to 6.9 billion assuming a 2% savings in care cost for six key medical conditions thought to most benefit; (d) additional savings in health care costs for private health care payers; and (e) royalty revenue of up to $1.1 billion from the intellectual property generated by the initiative. In addition, the initiative was proposed to create 5,000 to 22,000 new jobs per year over the funding period.

Regional competitiveness and the economic value of the activities took another turn when a decision was required to where the new stem cell agency would have its home. Five of the major cities in the state bid on the opportunity to house the agency, offering office space, hotel and conference space, commitment to provide development space near the agency and other inducements, each knowing that the presence of the agency would bring in significant private revenue - San Francisco was eventually selected as the site,

The agency has now become mired in debate before it can get off the ground. Groups seeking to declare the initiative illegal, to have more control of who makes funding decisions, to remove conflict of interest from the board members and to have more transparency in action have slowed progress of the agency, for the betterment, or not, of the state mission.

Since the initiative has passed, the U.S. administration has rethought their policy and more federal funding may be made available once controls are put in place for manipulation of the embryonic stem cells.

In 2004, researchers in South Korea announced they had cloned the first human cells. By the time that the announcements were proved false, one of the most public scientific frauds in recent history, a large amount of investment money had already been shifted in the world and opportunities for funding stem cell researchers elsewhere had been lost.

The saga will continue, but clearly it has already demonstrated that the desire for economic health and the competition between regions for attracting high-valued economic communities is a major driver for the fundamental research needed to provide long term public benefit.

Alternate Models of Investment

The models of investment described here have several basic tenets: that public health benefits from innovation in therapeutics, that innovation is high risk, that ownership of proprietary products is required for groups to be willing to take these risks, and that private enterprise is best suited to develop products. Certainly under this scheme tremendous public health benefits have been gained. Balancing these tenets with a time limit on ownership of intellectual properties has allowed generic products to make enormous in roads into the challenging health situation in the developing countries.

The model is not without its detractors, particularly in areas where profit and public health can find little common ground. Cost containment issues in Europe aside, there is little inherent motivation for private enterprises to develop therapeutics for malaria which affects a significant number of people in Africa and Asia each year, or other tropical diseases that have minimal impact on the public health of the U.S., Europe or Japan. This problem has not gone unnoticed even by the pharmaceutical manufacturers who recognize that a different model is required to address this enormous need *(87)*.

The United Nations' World Health Organization (WHO) has been the key governmental funding organization for such research and development for many years focusing their investments in developing countries that have experts in the diseases and problems themselves. WHO is often the major customer as well. Non-government organizations such as the International AIDS Vaccine Initiative (IAVI) have also been instrumental in funding such activities. They have supported research activities in the profit-driven hotbeds in developed countries and clinical studies in the developing world. The Institute for Oneworld Health is another example of a non-profit organization developing novel drugs or novel presentations of drugs that are applicable to use in the third world and that are of little financial value to large pharmaceutical companies.

In addition, the AIDS epidemic in Africa highlighted that rapidly emerging diseases provide a particularly poignant testimony to the dichotomy between the two worlds. In such cases, there are no generic drugs that provide any substantive relief from the problems and poorer countries suffer enormous public health crises in the face of the cost of the ethical drugs. This provides a significant moral dilemma for pharmaceutical companies and the developed world, however. Some of these drugs are key to the viability of the companies and the stances they have taken consequently have not had a positive effect on the public's view of the industry as a whole. The TRIPS agreement allowed for relief from intellectual property constraints during public health emergencies, but large pharmaceutical companies are worried about the slippery slope of such an exception and especially the ease of parallel trade once countries begin to manufacture products for in-country use. Success in the control and tracking of manufacturing sources and illegal importation should allow companies more comfort with this strategy though, hopefully to the benefit of all.

Because of the high cost of development, for-profit companies focus on products with potential for high returns, the "block buster" drugs for chronic treatments of cardiovascular disease, depression, and diabetes for example. Biologicals have shown themselves to be blockbusters when applied to small patient populations with life threatening or life debilitating diseases. But small patient populations that cannot return the development costs typically do not get attention from those that invest in innovation. These groups are typically ill served and alternate funding sources such as direct development by NIH or non-profit pharmaceutical companies become the key investors to develop these product. In most cases, these groups cannot afford the high cost of basic research and depend on available basic technology for development.

Others have been more expansive in their proposals to change the models of innovation and R&D suggesting that governments should do more to invest in product development themselves, absorbing the risk and providing the free right to market the drugs at a much lower cost to the public. This is probably the only answer for some diseases, however there are many obstacles to such an investment model no matter how much it appears to benefit the public good. Governments are typically poor at activities that require much flexibility, speed, and continual innovation and they are not good at providing the rewards needed to attract the most talented professionals to the discovery and development process. The U.S. model of giving money to other institutes to do research rather than running the research themselves as is done in other countries to a greater extent is probably one of the reasons for the effectiveness of the U.S. program.

However the demand for improved public health, the current dissatisfaction with the cost of healthcare and the transparency of global issues may nudge the current paradigms or allow them to accommodate alternate means of investing in pharmaceutical R&D and ultimately in public health.

Conclusions

Investment in pharmaceutical research and development is characterized by two major drivers in 2005. First is the concern by large pharmaceutical companies to fill their pipelines. Second is the competition among a growing number of biotech communities for enticement of the dollars that the large pharma companies are willing to spend to attain this goal and which the communities believe are critical to their overall economic health.

The United States continues to dominate in 2005 but communities are not complacent in their attention for continued position. There is a general concern in the science community about the ability to attract and offer exciting opportunities to the most innovative students and scientists.

Europe's biotech and pharma sectors are compromised by healthcare cost containment and the inability to provide a uniform regulatory, political and financial system. However, the national governments individually and collectively are committed to strategic improvements to increase competitiveness. Japan is in a similar situation, but although its pharmaceutical sector has matured its biotech industry is in its infancy. Government action to improve the research environment has been slow.

Asia is rising as a competitive threat to the older biotech communities and hopes their biotech sectors will become magnets for foreign investment. The breadth of activities is uneven, as is facility quality, and few countries have a sophisticated plan for successful development. Nevertheless, the region is changing too rapidly to prognosticate beyond the short term.

References

1. Industry Profile 2002. The Pharmaceutical Research and Manufacturers of America (PhRMA). http://www.phrma.org/
2. Kermani, F., Findlay, G. *The Pharmaceutical R&D Compendium: CMR International/Scrip's Complete Guide to Trends in R&D.* CMR International/SCRIP Publication http://www.cmr.org/ (accessed November 1, 2005)
3. Top 20 Pharmaceutical Companies Report, *Contract Pharma*, 2005, *7* (Jul/Aug), p 32
4. Reisch, M. Future Oriented Spending, *Chemical and Engineering News*, 2005, *83* (February 7), p 18
5. Top 10 Biopharmaceutical Companies Report, *Contract Pharma*, 2005, *7* (Jul/Aug), p 98
6. Schacht, W.H.; Thomas J.R. Patent law and its application to the pharmaceutical industry: an examination of the Drug Price

Competition and Patent Term Restoration Act of 1984 ("the Hatch-Waxman Act"), Congressional Research Report for Congress, January 10, 2005.
7. Jordan, G.E. Generic drug sales to soar past estimate, *Newark Star Ledger*, New Jersey, September 20, 2005.
8. Drews, J. *In quest of tomorrow's medicines*, Springer –Verlag, New York 1999
9. Gassmann, O.; Reepmeyer G.; von Zedtwitz, M. *Leading Pharmaceutical Innovation: Trends and Drivers for Growth in the Pharmaceutical Industry*, Springer–Verlag, New York, 2004; WGZ Brnachenanalyse Life Science. *WGZ Report*, October 2002, Dusseldorf
10. Incentives for Innovation: New Perspectives, 2004 Policy Conference, Merck Company Foundation, March 2004 www.merck.com/about/cr/pppi/incentives_for_innovation.pdf (accessed November 1, 2005)
11. See for example: Outlook 2005, Tufts Center for the Study of Drug Development, 2005 http://csdd.tufts.edu (accessed November 1, 2005)
12. Meeting of Health Ministers, Paris 13-14 May 2004 – Towards High Performing Health Systems., *OECD press release*. http://www.oecd.org (accessed November 1, 2005)
13. *The OECD Health Project. Towards High-Performing Health Systems, Summary Report.*, OECD, May 12, 2004 http://www.oecd.org/dataoecd/7/58/31785551.pdf (accessed October 24, 2005)
14. Schumock, G.T.; Walton, S.M. Expenditures for prescription drugs: too much or not enough? *Healthcare Financial Management.* October 1, 2003.
15. The Burden of Chronic Diseases and their Risk Factors, *National Center for Chronic Disease Prevention and Health Promotion.* http://www.cdc.gov/nccdphp/ (accessed October 24, 2005)
16. Lichtenberg, F. DTC Advertising and Public Health (power point presentation), 2003 frank.lichtenberg@columbia.edu
17. Statistics 2002: The pharmaceutical Industry in Germany; Facts & Figures. *Der Verband Forschender Arzneimittelhersteller (VFA)*, 2002 http://www.vfa.de
18. WHO (2003) Adherence to Long-Term Therapies. Evidence for Action, World Health Organization World Bank (2002) World Development Indicators 2002, CD-ROM Washington DC

19. *Laboratories of Innovation: State Bioscience Initiatives 2004*, Battelle Technology Partnership Practive and SSTI, June 2004, www.bio.org/local/battelle2004/ (accessed October 11, 2005)
20. *Beyond Borders – The global biotechnology report 2002*. Global Health Sciences, Ernst and Young, 2002.
21. *Pharmecutical R&D Outsourcing Strategies – An analysis of market drivers and resistors to 2010*. Reuters Business Insight. Healthcare, 2003
22. Population Division of the Department of Economic and Social Affairs of the United Nations Secretariat, World Population Prospects: The 2004 Revision and World Urbanization Prospects: The 2003 Revision, http://esa.un.org/unpp, (accessed 28 October 2005)
23. *The pharmaceutical industry in figures – key data, 2005 update*, European Federation of Pharmaceutical Industry Associations, 2005 http://www.efpia.org
24. *Biotechnology Industry Statistics*, Biotechnology Industry Organization, 2004 http://www.bio.org
25. *A survey of the use of biotechnology in the US industry*. US Department of Commerce. October 2003. http://www.technology.gov/reports.htm
26. Rovner, S.L. Academic R&D Spending Trends, *Chem & Eng News*, 2004, Nov 15, pg 37 and the National Science Foundation, WebCASPAR Database System.
27. *Report to Congress on Affordability of Inventions and Products*, Dept of Health and Human Services, Nathional Institutes of Health, 2004
28. Freeman, R.B.; Jin, E. (2003) Where do new US-trained science-engineering Ph.Ds come from? *AAAS*, 2003, Feb 13, www.ilr.cornell.edu/cheri/conf/chericonf2003/chericonf2003_03.pdf
29. Monastersky, R. Is there a science crisis? Maybe Not. *Chronicle of Higher Education*, 2004, July 9; http://chronicle.com (accessed October 11, 2005)
30. *Rising above the gathering storm – Energizing and employing America for a brighter economic future*, National Academies Press, Washington DC, 2005
31. Trumball, J.G. *Institutions and Industrial Performance: The Pharmaceutical Sector in France, Germany, Britain and the US. MIT IPC Working Paper 00-0002*. Massachusetts Institute of Technology Industrial Performance Center, 2000. http://web.mit.edu/ipc/www/pubwp.html

32. Kermani, F.. *Global Pharmaceutical Pricing: Strategic Issues and Practical Guidelines.* Urch Publishing, 2000. http://www.urchpublishing.com
33. *America's Biotechnology Report: Resurgence*, Ernst and Young, 2004.
34. Sikes, B.J. International genes that are the world's envy, *San Diego Metropolitan*, 2002 http://www.sandiegometro.com/2002/feb/coverstory.html
35. *MassBiotech 2010: achieving global leadership in the lifescience economy*, Massachusetts Biotechnology Council, Boston Consulting Group, 2002., www.bcg.com (accessed October 11, 2005)
36. *Biotech Ends the Year on a High Note, Quarterly Report*, Burrill and Company January 05, 2005
37. *The EU at a glance - The History of the European Union.* The European Union Online, 2005 http://europa.eu.int/abc/history/index_en.htm
38. *Biotechnology in Europe: 2005 comparative study, a report by Critical I for BioEuropa*, BioVision, Lyon, France April 13, 2005
39. *The Pharmaceutical Industry in Figures.* European Federation of Pharmaceutical Industry Associations, 2002 http://www.efpia.org
40. *Pharmaceutical Industry Statement on Conclusion of "G10" High-Level Group on Innovation and Provision of Medicines.* European Federation of Pharmaceutical Industry Associations. February 2002. http://www.efpia.org
41. *Innovation policy and performance: a cross-country comparison*, OECD Publishing, June 1, 2005, www.oecdbookshop.org, (accessed October 11, 2005)
42. Kermani, F.; Bonacossa, P. Putting it Together: How to Recruit and Maintain a Highly Skilled Clinical Staff. *Contract Pharma*. March 2003.pp1-5.
43. Kermani, F.; Gittins, R. Where will industry go to for its high calibre staff? *Journal of Commercial Biotechnology.* 2004 *11*(1). pp 63-71.
44. Woods, M. Europe slow in stemming 'brain drain' to America.. *The interactive version of the Pittsburgh Post Gazette.* October 20, 2003 http://www.post-gazette.com/pg/pp/03293/232608.stm
45. Henley, J. Scientists begin wave of protests by taking to streets. *The Guardian.* 10 March 2004. http://www.guardian.co.uk/france/story/0,11882,1165835,00.html
46. Kermani, F.; McGuire, S. Japan's ageing challenge. *TransPharma*, 2003 1(2), pp 34-37 http://www.chiltern.com/_data/Articles/16.pdf

47. *Facts & Statistics from the pharmaceutical industry.* The Association of the British Pharmaceutical Industry (ABPI), 2004. http://www.abpi.org.uk/information/default.asp
48. *UK Bioscience Industry Fast Facts.* Bioindustry Association, 2004. http://www.bioindustry.org
49. *Competitiveness and Performance Indicators 2002.* Pharmaceutical Industry Competitiveness Task Force (PICTF), 2002 http://www.advisorybodies.doh.gov.uk/pictf/pictfoneyearon.htm
50. *Bioscience 2015.* Bioscience Innovation and Growth Team (BIGT), 2004. http://www.bioindustry.org/bigtreport/
51. *UK sees healthy growth in investment, but European levels decline.* The Association of the British Pharmaceutical Industry, 2003. http://www.abpi.org.uk/press/press%20releases_03/030129.asp
52. *University confirms subject cuts.* BBC 20 December 2004. http://news.bbc.co.uk/1/hi/education/4105961.stm
53. *Select Committee on Science and Technology - Eighth Report.* Communications Directorate, United Kingdom Parliament, 2004, Westminster, London.
54. *Education system fails pharma industry.* Drug Researcher, Decision News Media, 2005. http://www.drugresearcher.com/news/ng.asp?id=58066-education-system-fails
55. *L'essentiel 2003.* Les entreprises du médicament (LEEM).2003 http://www.leem.org/
56. *OECD Health Data 2003 Show Health Expenditures at an All-time High.* Organisation for Economic Co-operation and Development. 2003, http://www.oecd.org
57. Masson, A. Médicament: *PharmaFrance 2004 - S'inspirer des politiques publiques étrangères d'attractivité pour l'industrie pharmaceutique innovante,* CERN .2004 http://www.cgm.org/rapports/publi.html
58. German Health Care Shake-up Draws Fire. *Deutsche Welle.* 22 July 2003. http://www.dw-world.de/english/0,3367,1432_A_925915,00.html
59. Rossiter, B. Higher mandatory rebates and price restrictions offer little incentive for companies to develop new drugs in Germany. *Med Ad News,* 2003 http://www.pharmalive.com/magazines/medad/view.cfm?articleID=244
60. *Germany's Pharmaceutical Industry.* Invest in Germany, 2004, http://www.invest-in-germany.de/en/

61. *Spain takes the bull by the horns.* IMS Health. October 2003. http://www.ims-global.com/
62. *The Pharmaceutical Industry in Figures.* Farmaindustria, 2002. http://www.farmaindustria.es
63. Lidden, J. The Global Pharmaceutical Revolution. *Business Facilities.* Aug 2003. http://www.facilitycity.com/busfac/bf_03_08_special1.asp
64. *The Pharmaceutical Industry in Spain.* Prepared for the General Directorate for Trade and Investment. Ministry of Economy. March 2002. http://www.investinspain.org/Pharmaceutical.htm
65. Desmet K.; Garcia E.C.; Kujal P.; Lobo F. *Implementing R&D policies: An analysis of Spain's Pharmaceutical Research Programme.* mimeo, Department of Economics. Universidad Carlos III de Madrid, 2003.
66. Kermani, F. Japanese R&D: Branching Out. *Applied Clinical Trials*, 2004, August 1. http://www.actmagazine.com
67. Kermani, F.; Gallagher, R. Japan: Settling the R&D jitters. *Good Clinical Practice Journal.* May 2005. http://www.gcpj.com
68. Murai, F. *Global Outsourcing Review*, 2003, *5(2)* pp 95-96 Elm Publishing UK.
69. *Biotech Japan*, Japan Biotechnology Industry, 2004 http://www.biojapan.org
70. *Hsinchu Science Park. Yearly statistics.* Administration of Hsinchu Science Park. http://SIPA/en/
71. *The Global Marketplace. Center for Industry Change.* Ernst & Young, 2000
72. Personal Communication, Ranbaxy Corporate Presentation January 2004 (Seema Ahuja, Senior Manager) Corporate Communications Department. Ranbaxy Laboratories Limited.
73. *R&D expenditure of the Indian pharmaceutical industry 1999 – 2003.* Organisation of Pharmaceutical Producers of India, Bombay, India, 2004 http://www.indiaoppi.com
74. Gwynne, P. Singapore – creating the Biolopis of Asia, *Science*, 2003 October 3 and http://sciencecareers.sciencemag.org/feature/advice/foc_10303.shl
75. Young, E. The game is on: Singapore overview. *NewScientist Jobs, Insider*, 2004 September 18, www.newscientistjobs.com/insider (accessed October 11, 2005)
76. Anon. The great fall of China? *The Economist.* 13 May 2004. http://www.economist.com/opinion/displayStory.cfm?story_id=2668015

77. WTO Entry Boosts China's Economy. *China Daily,* November 18, 2002. http://www.china.org.cn/english/49058.htm
78. Howard, K. Patent fights rumble in China. *Nature Reviews Drug Discovery.* 01 December 2004. http://www.nature.com/news/2004/041129/pf/nrd1595_pf.html
79. *USPTO Holds Seminar on Chinese Criminal Justice System for Intellectual Property Offenses.* The United States Patent and Trademark Office. 17 February 2005. http://www.uspto.gov/web/offices/com/speeches/05-14.htm
80. Jiahe, W. The internal and external environments facing the domestic pharmaceutical industry, July 2002. http://www.tdctrade.com/report/indprof/indprof_20703.htm
81. Jia, H. China moves to reform biotech policies, *Nature Biotechnology* 2004, pg 1004 , www.nature.com/news
82. Anon. Venture capital lukewarm to biotech sector, *People's Daily Online,* August 17, 2004 http://english.people.com.cn
83. Jia, H. Firms bite from outsourcing pie, *China Business Weekly,* July 28, 2004, p 8
84. *Stem cells – Therapies for the future?*, European Commission, 2002 http://europa.eu.int/index_en.htm
85. Research Councils announce £16.5m investment in stem cell research, MRC, 2004   http://www.mrc.ac.uk/index/public-interest/public-news_centre/public-press_office/public-press_releases_2004/public-27_may_2004.htm
86. Baker, L.; Deal, B. *Economic Impact Analysis of Proposition 71: California Research and Cures Initiative,* The Analysis Group, Inc, September 14, 2004, www.analysisgroup.com (accessed December 2004)
87. Azaïs, B.; Gajewski, M. *Research and development for neglected needs: lessons learned and remaining challenges,* IFPMA, 2004

Chapter 3

The China Challenge

Timothy C. Weckesser

President and CEO, Sino-Consulting, Inc., One Tower Bridge,
100 Front Street, Suite 1460, Conshohocken, PA 19428

Abstract

In the bulk of this paper we examine the larger market and economic challenges that China presents to the U.S. and the world: the *"per capita"* challenge, the entrepreneurial challenge, the multinational company challenge, and the market entry challenge. We discuss the incredible overall growth, the decisive role of foreign investment, the maturing supply chain, growing domestic quality and added value, and so on. We look at trends in several specific markets – automotive, telecommunications, software, and power - to illustrate the key points. Special attention is given to the *"per capita* challenge,"especially as it relates to the environment. We also look at the rise of the private sector and entrepreneurs in China, and the emergence of Chinese multinationals. In the remainder of the essay we discuss specific challenges related to market entry and doing business in China. In a postscript we discuss the "inward looking" nature of Chinese society.

The Big Picture: Economic Trends and Implications

The problem with writing an article about China is that it tends to be obsolete by the time it is published. This is, of course, because growth and change are occurring with such remarkable velocity – a velocity unequalled in history and in a country so vast that the consequences will be monumental no matter what they turn out to be.

Average annual GDP growth has been over 9% for the past 20 years! There are today nearly a half million foreign companies in China, up from virtually zero 20 years ago. The result is that in 2003 China became the world's number one target for foreign direct investment (FDI), surpassing the United States, by attracting over a billion dollars *per week* in 2004, and continuing apace in 2005. China's exports have grown from $13 billion in 1980 to $450 billion in 2003 (over half of which originate from foreign invested companies), a staggering rate of change.

What is somewhat surprising is that while China has been putting up these fantastic numbers for nearly a generation, it has not really made much front page news until very recently. We would mark the turning point around the middle of 2005 when, within a two-week period, Haier bid for Maytag and a Chinese oil company placed an $18 billion bid for Unocal of the U.S. These events seemed to get everyone's attention, especially the politicians', and China hit the cover page of *Business Week, Fortune, Foreign Affairs*, etc.

So the numbers are useful after all. Because if absolute numbers are changing too rapidly to track, the trends they indicate are not. In fact, the faster the absolute numbers become dated, the more the trends are underscored. In this essay we will try to speak on this level – on the level of trends and their implications – in the hope that our shelf life will indeed be a little longer than the norm, our relevance and utility a bit longer lasting.

With this in mind, we shall continue to illustrate China's remarkable rise to a world economic power.

According to the International Monetary Fund, 2005 GDP rankings will look like those shown in Table 1.[1]

There are many observations to be made about these numbers. First, at a 9% growth rate, China will more than *double* the size of its economy over the next ten years, which will easily catapult it into third place globally. Next, a few years later, China will likely overtake Japan. Let us say that surpassing Japan is even twenty years away. It is a quibbling

Table 1: 2004 Top 10 GDP Countries

Rank	Economy	Total GDP 2004 (in millions of U.S. dollars)
1	United States	12,438,873
2	Japan	4,799,061
3	Germany	2,906,658
4	United Kingdom	2,295,039
5	France	2,216,273
6	China	1,843,117
7	Italy	1,836,407
8	Spain	1,120,312
9	Canada	1,098,446
10	Russia	755,437
11	India	749,443

point. By the year 2025, it is hardly debatable that the two economic superpowers will be the U.S. and China.

Perhaps even more telling, if GDP is calculated on the basis of purchasing power parity, China is *today* number two only to the United States, with an amazing 12.6% of global output.[2]

A central, contrasting observation is that these absolute numbers belie a still very low per capita GDP rate - 26th place globally – and this is much harder for China to address. Although both the United Nations and the World Bank record that China has made "enormous progress" in reducing poverty and is in many ways ahead of its goals,[3] it is daunting to divide 1.3 billion into any number, no matter how rapid the growth.

The implications of these two factors – macroeconomic growth on an unprecedented scale and per capita parity growth on a microeconomic scale – is the critical dichotomy for China from which so many central problems *and* opportunities arise. We shall illustrate this as we progress.

The World's Workshop

For the last three years or so China's import and consumption of raw materials has strained global supply and driven up prices. China consumed about 8% of the world's oil, and between 20% and 40% of the world's aluminum, copper, tin, steel, cement, and other resources.

This binge seems to have peaked, according to a number of analysts, but the point here is *why* this import frenzy occurred. It came from growing Chinese companies, of course, but just as much from those 500,000 foreign companies we mentioned earlier. That is, prices have been driven up by three key elements, all interrelated:

- Infrastructure growth financed by the government
- Foreign companies producing for export, and
- Both foreign and domestic Chinese companies producing to participate in the phenomenal domestic expansion, from infrastructure to consumer goods.

The raw materials were imported to be deployed in numerous industries to support domestic growth, and much of these raw materials were also processed and re-exported to world markets. All this turned China into the "world's workshop" almost overnight. Table 2 shows some remarkable examples.

Table 2: China's Estimated Percentage of Global Production of Selected Products[4]

Commodity	% Global Production
Cameras	>50
TVs	30
Air conditioners	30
Washing machines	25
Refrigerators	20
Hard disk dives	37
Mobile phones	37
Digital cameras	50
Office equipment	28

Again, these production ratios are up from nearly nothing 15 to 20 years ago. Moreover, it is important to note that well over half of this production is by or on behalf of foreign-invested companies. It is the U.S., German, British, Japanese, etc. companies that have been driving the growth. When we look at the huge trade imbalances, it is helpful to remind ourselves that a significant share of our imports are from our own companies that are making products in China and exporting back to the U.S. This fact can be viewed from many different angles – employment shifts, technology transfer, global competitiveness, quarterly share value, etc. In any case, it is certain that, from a macroeconomic view, companies set up production in China because they must do so to remain competitive.

For example, China is Motorola's global production base for cell phones. If Motorola had not made this commitment, it could easily be argued that the company simply would not be a player today, that its share value would be on an irreversible skid, and that the U.S. layoffs required to survive would be devastating.

Let's now look at just a few specific industries. They are enough to illustrate the point.

Automobiles

In 1999, there were 220,000 vehicles sold in all China; in 2004, the number was over two million – almost a ten fold increase in just a five year period. From 2000 to 2003 – just three years - the number of privately owned cars doubled from five to 10 million. Projections for privately owned cars are roughly as follows:

- 2010:   7.8 million
- 2020:   16.8 million
- 2030:   45 million

All of the major players are there with one or more joint ventures, and they are doing well, despite current reductions in prices and profits because everyone has joined the game. (Automotive is still a "protected" sector in China where joint ventures are required.) They have joined the game because the automobile penetration rate is so low and has nowhere to go but up. If competition is bringing down prices, well, that's normal market evolution, but the long term prospects are still the best in the world.

The growth has been so rapid that today China is the world's 4th largest auto producer, after the usual line up of the U.S., Japan and Germany. Germany's Volkswagen was the first big foreign player in

China, and has been well-rewarded for its risk. In 2004, Volkswagen sold more cars in China than in Germany – the first time in its corporate existence that sales in a foreign country outstripped the home market. In 2005, however, latecomer GM overtook VW to capture number one market share position. (Whatever GM's woes in North America, it is doing great in China.) Also in 2005, Honda began to export complete cars from China, a trend that is spreading to the other OEMs (original equipment manufacturers). From this small export start, China will become a global leader in vehicle exports over the next ten years.

Figure 1 shows market share distribution in the automotive market in 2004.

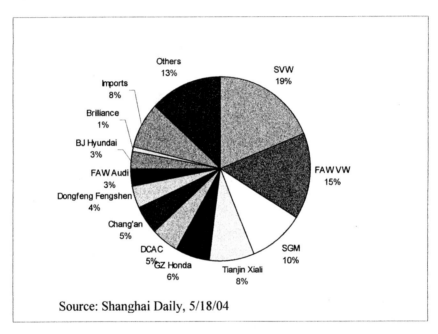

Figure 1: Automotive Market Share in China

While Volkswagen's two joint ventures have been dominant, it can be seen that GM had already picked up about a 10% market share - despite the fact that it was a quite a latecomer to the market. Today, you will see a rather high-end version of the Buick Regal everywhere on the streets of Shanghai.

Another interesting point is the proportion of cars that are still imported. These are all quite high-end, expensive vehicles, made even more expensive by high tariffs and the value added tax. Still, there is a strong market for them among emerging rich business people in China and foreign executives.

All this new manufacturing capability has pulled the vehicle parts suppliers into China as well, and foreign companies – Visteon, Delphi, Mahle Group, Bosch – heavily dominate the market.

Very importantly, pressured by government targets of 70% local content within three years, OEMs have deliberately pressured their suppliers to set up production facilities in China to bring the supply chain next door and reduce costs – a pattern common in many industries. Based on current rules, if vehicles have more than 40% imported components, they are subject to tariffs as "complete vehicles," which makes them much more expensive than locally made vehicles. Local content is therefore a must in an increasingly competitive environment. This in turn pulls foreign suppliers into China if they wish to get a piece of this growth market while it is still young.

Motorola, in the electronics market, is another example. Its regional procurement policy is clear: you have to be in China (to bring down your own costs) if you want to continue to be a supplier. In the meantime, they have identified and groomed numerous indigenous Chinese suppliers. This important market pull phenomenon will be discussed again later.

Telecommunications

Figure 2 illustrates the fantastic growth of China's telecom market between 1996 and 2004. You will note the small number of users just nine years ago in 1996, and if we went back to 1990 the numbers are negligible. Today, when a new building goes up in China, it is wired with optical cable for broadband Internet and telephone service. All hotels offer high-speed wireless Internet connections – long before their American counterparts – at nominal cost and often free.

For a number of years, new wireless subscribers have been increasing at the rate of over a million *per week*. This trend continued in 2005. China Mobile and China Unicom added over 28 million new users in the first half of 2005. While this represents a small slowing of the growth rate, it has not yet significantly hit revenues. Revenues are up

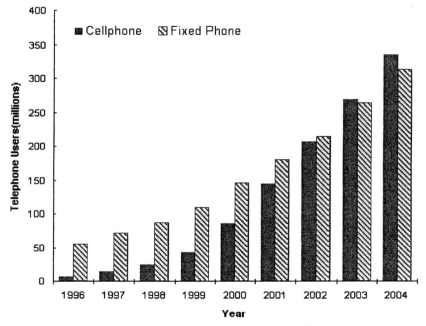

Figure 2: Fixed and Mobile Phone Growth in China

over 10% year on year. All told, China has more cell phone subscribers than the entire population of the United States, with around 350 million and growing.

Not surprisingly, therefore, China is today both the largest consumer and producer of cell phones in the world. According to CCID,[i] a market research company in the IT industry, handset sales will reach about 90 million units in 2005, which would be more than a 25% increase over 2004. Although domestic players like Bird and Konka have done a great job getting into the ballgame, the market is still dominated by foreign players – Nokia, Motorola, and Samsung. Handset production capacity in China is said now to be about 500 million handsets, up from about 150 million just four years ago (2001). Again, the entire supply chain is solidly in place.

In terms of infrastructure to accommodate this growth, carriers try to stay at least six months ahead of the curve. In addition to the big

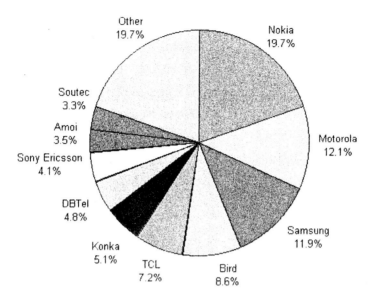

1. Percentage is based on handset sales in China (excluding exports), including GSM/GPRS/CDMA. 2. Companies with less than 3% market share include Kyocera, LG, NEC, Eastcom, Capitel, CECT, Haier, Hisense, Kejian, Panda, ZTE and others.

Source: Chinex, from MII, company reports.

Figure 3: Handset Market Shares

multinationals in this field, China has bred its own giants, especially Huawei and ZTE, the former wholly private and the latter state-owned. Both companies are less than 20 years old and are now multi-billion dollar enterprises with significant global sales. Huawei has 55 offices around the world with over 40% of it's over $5 bilon in sales coming from non-Chinese markets. Its international sales have doubled every year for the last 5 years, and they are very proud of this. ZTE is by far the largest Chinese IT company with sales over $20 billion, some 20% of which is from foreign markets.

A critical observation in all of this is how China is rapidly moving up the value chain. For example, China has had the reputation of being poor at design but good at manufacturing. That seems to be rapidly changing. Domestic headset companies are investing heavily in local design, and even the leader, Nokia, is increasingly hiring Chinese designers. In fact, one of Nokia's best selling phones in China was completely developed by local talent.

A similar trend is occurring with integrated circuits. China has been an enormous consumer of ICs, accounting for some 20% of global consumption. However, over 80% of these chips are imported. China would like to change this, but IC design and testing is very expensive, not a project for your local garage entrepreneur. One effort to address this is the Shanghai Research Center for IC Design.[6] The Center has received investment from the Science and Technology Ministry and Shanghai city and is a kind of industry-specific incubator. What they did was simple in concept: they bought the very expensive software and test equipment and made it available to design entrepreneurs who could otherwise never afford such technology, and they provide training for would-be entrepreneurs and existing Chinese design houses. Moreover, they have partnered with giants like LSI Logic and Chartered Semiconductor Manufacturing. The upper floors house the equipment and design rooms, and the lower floors house the companies that grow out of the top floors - and it is packed. Of course, like most incubators, they have numerous failures. Yet there are also successes and, equally important, the training creates a foundation for the future.

This movement up the value chain is critically important. China has already set a goal of 5% of global IC production by 2010. They may well not reach that goal, but the intent and the strong efforts behind the intent make the point. As Thomas Friedman, author of the popular *The World is Flat*, has said, "So in 30 years we have gone from 'sold in China' to 'made in China' to 'designed in China' to 'dreamed up in China.'"[7] Imagine the rise of Japan in the 1970s and 1980s, but multiply by 10.

Software

China's software industry has also seen explosive growth. Table 3 shows this growth, broken down by software products, services, system integration (SI), and export, and you will note the last row that shows huge year-on-year expansion rates that are increasing rather than slowing.

Compared to the automobile and telecom sectors, domestic Chinese software companies have fared much better than foreign companies in China. The government provides strong and direct support to development of the domestic software industry.[8] There are essentially

Table 3: Value of Software Industry in China 2000-2004 (million U.S.$)

	2000	2001	2002	2003	2004
Software Products	2,878	3,990	6,135	8,464	11,125
Software Services	3,894	4,909	5,666	2,479	5,042
System Integration	*	*	*	6,409	9,045
Software export	399	726	1,499	1,995	2,600
Total Industry Value	7,170	9,625	13,301	19,347	27,812
Growth of total value	34%	34%	38%	45%	44%

Source: Annual Report of China Software Industry 2005[9]
* Included in software services.

two strategic targets for its policies: one is to strongly encourage the development and deployment of open source software (OSS), such as Linux; the second is to reserve the huge realm of government digitalization for domestic companies. Because Beijing is investing heavily in the concept of 'e-government', this means big money for domestic enterprises that can rise to the occasion.

Since these policies are directly antithetical to Microsoft's proprietary OS, the software giant has had consistent problems in China and, in general, has not done well. Certain other foreign companies have done very well in China. For example, in recent years, the management software market has come on strong. According to CCW Research,[10] the size of the overall management software market in China reached U.S.$1.75 billion in 2004, 28% up from 2003. The growth rate in 2005 is projected to be another 28%. Such giants as SAP and Oracle have captured a significant share of this growth, but the leaders remain indigenous companies like UFSoft and Kingdee.[11] Moreover, SAP and Oracle have only been able to capitalize on the enterprise market, not the government market.

It is instructive to summarize the government's support measures for the software industry:

- Ministry of Information Industry (MII) has established a Working Office to implement key policies. The office has nine working groups and significant staffing.
- The government is taking the lead by using Linux in its e-government initiative and encouraging provincial and local governments to do the same
- The Government Procurement Law gives clear preferential treatment to small and medium-sized enterprises, and sets aside government procurement for domestic companies and products that meet environmental protection requirements. (One key reason for this is security. It is virtually impossible for the Chinese government to depend on a foreign company's software for its safety.)
- MII and 20 major Chinese enterprises and organizations - including the China Software Industry Association, the China Information Industry Trade Association, Co-Create Open Source Software Co., Ltd. and Red Flag Software Co., Ltd., etc. - have established the China Open Source Software Promotion Union (COPU) to rationalize the development of open source software in China and promote cooperation in this effort in all of Northeast Asia.
- MII also encourages the formulation of industry and national standards by competent enterprises. Several already developed include the Chinese Linux Application Programming Interface (API) Norm, the Chinese Linux Desktop OS Technology Norm, the Chinese Linux Server OS Technology Norm, and the Chinese Linux User Interface Norm.
- The government also supports the transformation of R&D organizations into enterprises, that is, the transformation from governmental or academic status to private enterprise. (In other words, entrepreneurship.)
- There is also a move to establish, finance, and cultivate "backbone" software enterprises. MII and the former national planning committee have jointly stipulated *Management Measures of State Software Industry Bases* and approved 11 key software parks as state software industry bases.

- The government has funded numerous special software projects, such as the *Electronics Information Industry Development Fund* under MII, the *"Double Program" project* under MII and the China Bank of Industry and Commerce, a number of software projects under the famous *863 Program* run by MII and the National Defense Science and Engineering Committee, projects under the *Torch Program* managed by MOST, the *Innovation Fund for Science and Technology Type Small and Medium-sized Enterprises,* and others.

Perhaps the biggest problem in the software sector is piracy. Importantly, it is a problem for both foreign and domestic software developers. An IDC survey claimed that while 36% of computer software worldwide is pirated, the figure in China is 93%, reflecting a total loss of U.S.$3.8 billion in 2003.[12] China debates these figures, but readily concedes the general magnitude of the problem.

Importantly, piracy is heavily dependent on which software sector you are in. It is generally agreed that the real problem is with PC software, whereas it is a fairly small problem in specialized software service and enterprise application software. For example, there is little piracy in highly specialized service software such as is developed by Accenture, IBM, PricewaterhouseCoopers and EDS or in enterprise software such as is developed by Oracle, SAP, CA and UFSoft. Rather, Microsoft and Adobe products are typically the types of software that are pirated.

We will address the issue of IP protection again later.

It is clear that Beijing has made strategic decisions to promote the development of China's software industry in general and OSS in particular. Support has been continuous and steady in terms of policies and a substantial supply of capital. In spite of this, however, the Chinese OSS industry is still at an early stage of development, and users still do not have great confidence in OSS producers and their technology.

More broadly, the domestic industry is almost certain to command the lion's share of all government business, while foreign companies will still have good opportunities in the enterprise sector. Even in this arena foreign companies will be challenged by increasingly competent domestic players. Price will continue to be a key variable, but price will be viewed in the context of the total package being offered. Chinese enterprises are increasingly seeking *total solutions* that can enhance profitability, not just software. That is, they do not want to buy software; they want to buy a path to better profits – just like the West.

Power and Pollution

With the kind of growth we have been describing it is little wonder that electricity supply has not kept pace with demand. It is one thing to stay six months ahead of the development curve in wireless telecom as base stations are relatively quick and easy to install. (Of course, technology changes and upgrades add their own level of complexity in this industry.) It is quite another to try to stay six months or a year ahead of the energy curve. Power is a national strategic industry. Building a power plant requires cooperation among the power company (there are six of them which cover all of China), the central government, the regional and municipal governments, the influential electricity research institutes, foreign and domestic suppliers, and others.

Figure 4 shows the growth of installed capacity from 1997 through 2004. The top line shows year-on-year percentage increases in capacity.

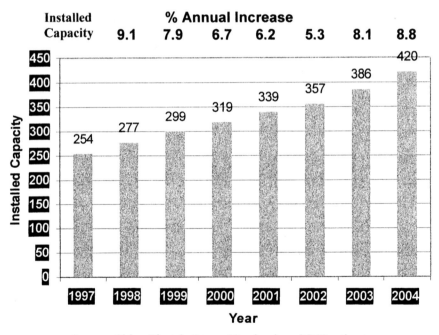

Source: China Electric Power Yearbook and SCI estimates

Figure 4: Installed Generation Capacity (1,000MW)

Throughout the 1990s the power supply was adequate and often in oversupply, despite consistent double digit annual growth. However, with nearly the same growth rate after 2000 on a much larger accumulated base, problems quickly emerged. Since 2002 China has experienced many power shortages, sometimes severe and widespread. Companies in many areas are still required to stagger their work weeks and planned brown-outs are common. The latest five-year plan goals, for 2001-2005, have been repeatedly revised upward.

Today, both China's installed capacity and output are 2nd only to the U.S. in the world. In Asia, China now accounts for over half of all energy consumption outside of Japan. Still, new installations and the upgrading of old ones continue apace.

It is likely that supply will catch up with demand within the next few years and most shortages will be limited. Still, this does not mean that there will be a "glut" of energy. This is because, again, on a *per capita* basis, China has a long, long way to go. Economic growth will almost certainly slow, and even slow considerably, but the hopes and dreams of individual, hard working Chinese will not slow. Indeed, they will very likely accelerate as a middle class standard of living becomes more tantalizing, more accessible, and more visible everywhere.

Figure 5 illustrates this issue well. It shows relative *per capita* consumption of several industrialized countries compared to China. The issue is obvious. China's averaged individual consumption is less than a third that of Japan, Taiwan and South Korea, and *less than a sixth of U.S. consumption*!

Where does this lead? What if China reached the U.S. level of energy consumption? Here is food for thought: it has been estimated that if China reached the consumption level of North America, she would require 80 million barrels of oil per day – equal to the entire current global daily consumption!

Of course, this is not possible. It cannot come true, so to speak. Now, we will not argue here the degree to which this extrapolation is balanced by consideration of other currently available energy sources. For our purposes it doesn't matter. The value of such extrapolations is not what they tell us about the future but what they tell us about the present. Put differently, and a little more dramatically, they are valuable because they tell us what we must change if we are to survive.

China knows this well. This knowledge leads us both to great challenges and great opportunities for at least the next 20-30 years.

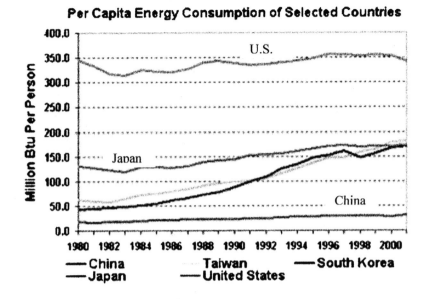

Source: U.S. Energy Information Administration

Figure 5: Per Capita Energy Consumption

We now look at the various sources of energy currently employed and in the future in China and consider the implications. Figure 6 shows the present distribution of power sources in China, and Table 4 shows projected sources.

It seems that every estimate one comes across for current and projected energy sources varies, sometimes widely. It depends on who's doing the reporting and what the assumptions are. However, there is one thing everyone agrees on – coal is going to be the dominant source of energy for a long time to come.

The data above shows that coal accounts for over two-thirds of all sources currently, and when we jump out 25 years to 2030 it drops slightly, to a projected 62%. Then in 2050, 45 years out, the ratio is projected to drop to 35%.

This is all encouraging, but it is quite long term and subject to many questions and unexpected turns of events. Even the small drop over the next 25 years may just be wishful thinking, or the government's "face" to the world. Yet the numbers may also be feasible. The government is sincere in its desire to change the ratios and clean the environment, and it

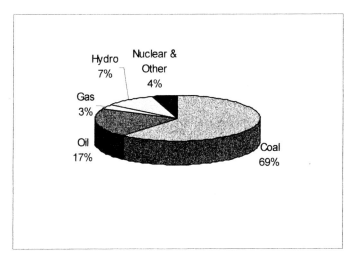

Source: State Development and Planning Committee

Figure 6: Energy Sources in China

Table 4: Projected Sources of Energy in China

	2004	2030	2050
Coal	69%	62%	35%
Oil	17	18	40-50
Gas	3	8	
Hydro	7	9	
Nuclear	2	3	15-20
Other	2	?	

Source: State Development & Planning Committee

is devoting large sums of money to this end. It is equally committed to sustained economic growth, and these two goals are not necessarily compatible. In fact, they can be in direct conflict. It is a tough situation.

The fact is that China has the largest coal reserves in the world and it can't afford *not* to exploit them. Today China is the world's largest producer and consumer of coal, and the second largest coal exporter. Thus while coal's share of overall Chinese energy production is projected to fall – and may indeed do so - coal consumption in absolute terms will actually *increase* - at least 5% per year it is estimated. Put differently, the power pie is getting larger, as simply illustrated in Figure 7, below. Thus a smaller percentage can actually represent substantially larger absolute consumption.

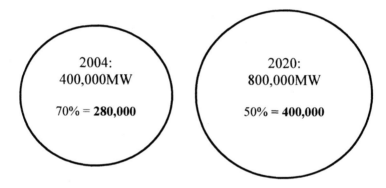

Figure 7: Simulated Growth in Total Energy Consumption

This dependence on coal puts tremendous pressure on the coal mining industry to step up production and on the railroads to transport the coal. (In fact, transportation has been the biggest bottleneck in the system.) In the future China will be producing much more than its current 1.7 billion tons of coal a year. Production is increasing at about 15% per year. This means, in turn, increasing pressure on mines to produce, logistics problems, corresponding issues of human safety, and a choking pollution problem.

The safety problems are staggering. In each of the last two years (2003, 2004) around 6000 lives were lost in mining accidents in China, and this number is on track in 2005. In the U.S., if there is a single coal mining fatality, it is cause for national front page news. In China, an average of fifteen fatalities occurs *every day!*

China produces one-third of the world's coal, yet has over 80 percent of the fatal accidents in the world's coal mining industry. There are some 28,000 coal mines in China, and 24,000 of them are small, private ones which produce about one-third of the coal mined, but account for the great majority of accidents. Of those killed, only about 28% worked in the large, state-owned mines.

The Chinese government is very serious about trying to address this terrible problem of mine safety. However, it is difficult to address for many reasons. Government has considerable control over safety standards in large state-owed mines, but virtually no control over the small private mines where most fatal accidents occur. In 2000, the government set up a national surveillance system to keep a close eye on the safety conditions of coal mines. They earmarked over 4 billion yuan (over U.S.$480 million) to help state-owned and small local mines in gas explosion prevention and monitoring. They also set safety goals which aim ultimately at reducing the national fatality rate per million tons of coal to about 0.4 by 2020, which would be similar to more developed countries.

Addressing the larger issue of heavy dependence on coal, China is attacking in many directions. For example, she is investing heavily in clean sources such as natural gas and hydropower. The huge Three Gorges hydropower project is well known, and another major project is being developed along the Yellow River that will be nearly as large as Three Gorges. At the same time, although the ratio of natural gas consumption will remain relatively small, the absolute numbers will grow dramatically, as illustrated in Figure 8.

Likewise, gas turbine production is projected to rise rapidly, as shown in Figure 9. Reaching this goal has become more feasible with the discovery of large gas reserves in Western provinces, and subsequent investment in a monumental "West-to-East" gas pipeline and the signing of huge gas import contracts.

The story with nuclear power is similar. China plans to have 27 new reactors in operation by 2020 producing some 36GW of power. Still, even when fully deployed they will account for less than 5% of total installed capacity.

Further, China is investing in renewable energy sources and has set a goal of producing 10% of needed energy from hydro, wind, and solar sources by 2020. Wind farms alone are supposed to generate 20,000MW of power. This figure is actually mandated in the Renewable Energy Law passed earlier this year.

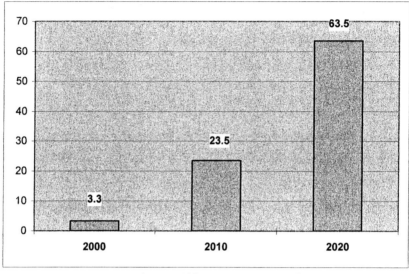

Source: former China Power Corporation

Figure 8: Projected Natural Gas Consumption, billion M^3

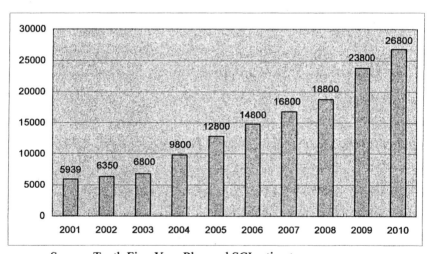

Source: Tenth Five-Year Plan and SCI estimates

Figure 9: Projected Installed Capacity of Gas Turbine Power Plants in China - 2001—2010 (in MW)

Still, with all this investment in cleaner sources of fuel, coal-fired power in 2020 is projected to be *twice* what it is in 2005 in absolute terms.[ii] As for the inherent problems of coal itself, China has outlined a ten point program for energy efficiency in its eleventh five-year plan (2006-2010). The points include:

- Upgrade coal burning industrial boilers, most of which are badly outdated, terribly inefficient, and polluting. This includes shutting down some plants, renovating others, and emphasizing use of clean technologies, especially circulating fluidized bed (CFB) boilers. The aim is to raise efficiency of existing coal burning units by just a few percentage points, which will translate into savings of 35 millions tons of coal

- Cogeneration, to raise heat efficiency, which will result in savings of 35 million tons of coal.

- Saving or replacing 38 million tons of oil

- Energy conservation for buildings, reducing consumption by 50%

- Environmentally friendly lighting which can reduce consumption by 70-80%

And so on... One could cite other investments as well. For example, the first small-scale projects for coal gasification have been undertaken. China aims to reduce sulfur-dioxide emissions by 50% by 2010. She has also adopted legal and political incentives such as taxes on high-sulfur coal, "coal-free zones" (e.g. around Beijing in preparation for the 2008 Olympics), and special taxes on older and more polluting producers.

The Environmental Challenge

All of these measures are encouraging. Certainly they are pointing in the right direction, but they may not be nearly enough. It can be cogently argued that the single greatest challenge that China faces is monumental environmental degradation and resulting pollution. Moreover, the effects reach far beyond China's borders, the most immediate victims being Japan and Korea. Atmospheric and water degradation have a global impact.

According to the World Health Organization, seven of the world's ten most polluted cities are in China. The largest contributors are industrial boilers and furnaces, as hinted above, which consume almost half of China's coal, although vehicle pollution has become a tough rival, as discussed below.

These problems are born of rapid industrial growth and by extremely rapid urbanization. Today 37% of the population lives in urban areas, up from just 13% in 1953. China's struggles with severe air pollution and acid rain, polluted lakes and rivers, river flow cessation, deforestation (which go back at least to Mao's "Great Leap Forward" in the late '50s.), and so on, are seemingly endless, and seemingly impossible to make a dent in as the country is undergoing such dramatic industrialization. Jared Diamond, in his recent book *Collapse*, devotes an entire chapter just to the subject of China and the environment. He points out that "China is already the largest contributor of sulfur dioxides, chlorofluorocarbons..., and (soon) carbon dioxide to the atmosphere." China has only 0.3 acres of forest per person compared with the world average of 1.6 acres. Overall only 16% of her land is forested, compared, for example, to 74% in Japan.[14]

A key issue in atmospheric pollution is the problem of inefficiency. To generate $1 of GDP, China uses nearly five times more energy than the U.S., nearly eight times more than Germany and over eleven times more than Japan. Coal consumption per kilowatt hour is 22% higher than the U.S. Power wastage in China is some 47% more than Western countries. (It should be noted that the problem is as much with the grid – transmission – as with production.) This inefficiency is a huge issue.

Enforcement of existing environmental rules and standards is also a problem. The rules and standards themselves are unquestionably progressive and, as noted earlier, pointing in the right direction. China is a big place, however, with many political and geographical jurisdictions, each with its own sociopolitical and economic relationships ("guanxi") that still, even today, can outweigh the law. (In addition, the law is always written in such a way as to leave much to interpretation in a rapidly evolving reality.)

Moreover, it seems that as soon as China begins to address one environmental problem, another emerges that completely counterbalances any progress. The prime example right now is coal and cars. In 1986, during this author's first visit to China, coal dust was so thick in the air you could almost taste it. You could most certainly see it on your clothes and feel it on your flesh as each day wore on. The coal dust was from inefficient utilities and even more inefficient industrial boilers. Today, much of this visible coal dust pollution has been cleaned up in the major Eastern cities because of the use of more efficient burning technology and aggressive clean-up policies for industrial boilers. Nevertheless, anyone who travels to China today will tell you the situation is still very poor. Moreover, it is now estimated that in the major cities cars have replaced coal as the dominant source of pollution – an insulting reward for 15 years of environmentally friendly legislation.

How will all of this end up – for China and for the world? We do not know! Credit Suisse experts believe that the economic pressures of her environmental policies will lead to "structural change" which will increase use of natural gas (to 7%) and nuclear (to 4%) by 2020, while coal will decline from 68% to 64%, and pollution taxes may drive the worst offenders out of business after the shortage of recent years passes.[15]

Perhaps, but this kind of thinking tends to be guided by the paradigms of the past. It is the "fossil fuel paradigm," so to speak. China in fact has a wonderful opportunity to lead the way in the creation and application of a new paradigm. Just as China, in a sense, leapfrogged wire line telecommunications to become the biggest wireless market in the world almost overnight, so too she can take up the challenge and lead the world in creating a new energy paradigm.

There is no doubt that China must still work within the paradigms she has inherited from the rest of the world. Some issues – such as inefficiency – can be addressed effectively now with technology and money, and China does indeed have huge foreign exchange reserves. However, can enough be done in time? Or to put it in a little scarier context, is it *possible* to do enough and in time? It is the old *per capita* issue again.

Earlier we mentioned what would happen if China reached *per capita* parity with the U.S. in oil consumption. The same kind of extrapolation can be made in numerous areas. According to some extrapolations of the Earth Policy Institute (EPI), based on real projected GDP growth of around 8%, China's economy would double in size every nine years, and *per capita* income would rise to about $38,000 by 2031, equal to the U.S. in 2004. If we extend this extrapolation to other areas, here's what the EPI comes up with for the year 2031:

- China's grain consumption would rise to 1,352 million tons, or about two-thirds of all the grain harvested in the world in 2004

- China's meat consumption would rise to 181 million tons, or about 80% of current global production

- China's coal consumption would be 2.8 billion tons, 0.3 million tons greater than total world production today

- China would have over a billion cars, compared to a global total of about 800 million currently.[16]

We could go on.... The question of "if" is even more pressing because India and other Asian nations are rising to the China challenge. Hundreds of millions of people in China and all of Asia are dreaming of becoming middle class citizens and, even more important to them, assuring a slot in the middle market for their children.

Yet the reality is that these extrapolations are impossible. Can China change the paradigm? Does she have the will? We do know that when China makes up its mind to do something, it gets done. It is a cliché to point to the Great Wall in this regard, but there are endless examples. The gigantic Great Hall of the People was built in under a year under Mao, a project that would take multiple years in any other culture. Millions of people were displaced to make space for the Three Gorges hydroelectric project and keep it on schedule. Most relevant here is the fantastic industrial development we have explored. This was not simply an accident of historical time or place. It was *created* by the most progressive policies in Asia. (Compare, for example, rules on equity ownership in Malaysia that still requires strong indigenous equity participation, while India has liberalized significantly but still lags behind China.) If you can get 1.3 billion people all looking in the same direction, you can accomplish a lot. China did it with such efforts as the one child policy, but the energy *per capita* challenge exhausts the imagination.

Rise of the Private Sector

Many people still think of China as a "communist" country in the Cold War sense of the word, but it is a misnomer. It is true that China is a one-party system and that party is still the communist party, but the word "communist" is like a remnant of the past that for many reasons cannot be abandoned. The economic reality is far more capitalist than socialist and there is hardly anything left of the old Marxist/Leninist ideology. The drive is toward a "market-oriented economy" and this has meant a shift to the private sector. This shift has profound consequences.

The numbers are quite clear in this matter and the trend unmistakable. A few words on the following tables may still be appropriate. Table 5 is derived from the China Statistical Yearbook 2002 and Table 6 is derived from the China Statistical Yearbook 2004. It is often difficult to get the same data on a given subject from two different sources in China. Methodologies are still evolving, so one can easily run into head-scratching inconsistencies when trying to find reliable numbers. As we discussed at the beginning of this essay, one can much more easily see trends even if absolute numbers do not agree.

Table 5 shows a dramatic shift to the private sector in the 22 years from China's "opening up" to 2002, from virtually zero contribution of private companies in 1980 to over 50% in 2002. It can quickly be argued that many so-called "public" companies are state-owned enterprises in disguise, as they are still largely controlled by the state.[17] It is true. At the same time, it is also true that the Chinese government has moved aggressively to divest its holdings in non-critical industries, and the general trend is strong.

Table 5: Percentage GDP Contribution by Type of Company in China, 1980-2002

	1980	1990	2002
FIEs*	~	<1	18
Public	1%	4 9%	19 54
Private		5	17
Collective	24	36	21
State-owned	75	55	25

Source: China Statistical Yearbook, 2002
* Foreign-invested companies

Foreign invested companies (FIEs) have grown from zero to over 20% of GDP today (18% in 2002 in Table 5), and today *over 50% of export value is contributed by FIEs*. (We need to remember this when we rant about Chinese exports. Over half of those exports are by American and other foreign companies making things in China and shipping them back here for sale.) This was not an accident of the "opening up," but the result of an aggressive investment incentive regime begun in the '80s. Policies originated at the national level and received countless enhancements and twists from various provinces and municipalities throughout the country, often beyond the control or even interest of Beijing.

China's nationwide blossoming of 'Special Economic Zones' (SEZs) and 'Economic and Technological Development Areas' (ETDAs) was at the core of this investment success. They became the physical focus for attracting manufacturing and new technology by offering strong tax incentives, tariff import waivers for production equipment, rebates on value added taxes, re-investment tax incentives, special "high tech" tax incentives, etc.

In fact, these policies created a more favorable business environment for FIEs than for domestic Chinese companies, a seeming irony. It seems that Beijing knew that, in fact, domestic enterprises were woefully lagging behind world standards, and that the best way to bring them along was to attract the best in the world to teach them. They of course had a great carrot to do just that – a market of over a billion people and an inexpensive, eager, and smart workforce. Unlike other countries (e.g. India) that insisted on protection for its domestic companies and equity ratios that favored domestic control, China opened its door rather completely in order to *learn* – about technology, about modern management, about production processes, and so on.

The bottom line is that these strategic incentives, launched over 20 years ago, have been remarkably effective, as the data show. In fact, they have been so effective that Beijing is now moving to level the playing field for domestic enterprises. It is time. It is fair.

Table 6 adds to this perspective by looking at employment in various sectors. (One can forgive the rather unusual time frame of 11 years, from 1992 through 2003.) What really pops out, of course, is the dramatic rise in employment – nearly five fold - in the purely private sector.

TVE ("township and village enterprise") employment has also increased, but very modestly at just 28% over the entire period. Moreover, these enterprises can be somewhat elusive and hard to define. Growing out of old collective enterprises, they come in many sizes and

Table 6: Employment Distribution, 1992-2003 (millions)

Employment category	1992	2003	% Change
Private companies	8.4	49.2	486%
TVEs*	106.3	135.7	28%
Total private & TVE	114.7	184.9	61%
State-owned	108.9	68.8	(37%)

Source: China Statistical Yearbook, 2004
*Township & village enterprises

shapes. Many are much closer to what we consider to be private enterprise than to state-owned enterprises.

At the same time, employment in state-owned enterprises has declined sharply, 37%.

Beijing hoped that its FIE policies and support of private enterprise would help to pick up the unemployed that were sure to follow from the decline of non-competitive state-owned companies. If we can judge by these data, it seems they have met with at least partial success on this score as well. Total private sector and TVE employment in 2003 exceeded employment losses in the state-owned sector by some 30 million, not a trivial number even in China.

In Table 5 we showed a category called "Private," which is separate from FIEs and public companies. These are largely domestic Chinese companies, mostly entrepreneurial and mostly family or individually owned. (In Table 5, these companies would be embedded in the first category, "Private companies.") This is a most interesting category and deserves special attention.

The Entrepreneurial Challenge

It is estimated that domestic Chinese private companies now account for some 30% of China's GDP. If we add to that the more than 20% from FIEs, and probably another 20% from TVEs, we are looking at an economy already powerfully dominated by non-state-owned enterprises.

We should not be misled into thinking that the government role has somehow diminished as a result of this trend. It has not. It has *caused* it and therefore still has the power of the creator, so to speak. Although it is highly unlikely that Beijing would ever try to undo its creation, no investor should ever make the mistake of thinking that the government's law now has more power than the government itself. Notwithstanding this caveat, the government itself has encouraged entrepreneurship – just keep your nose clean as you pursue your dream.

America has long been the world's bellwether in entrepreneurship. It has been a point of extreme pride for the U.S. and we have been a global model regarding how to nurture and develop innovation, new company formation, and development. To this author, it is the essence of America's inherent competitive edge. The entire concept of "business incubation" developed in the U.S. and spread throughout the world based on sound economic and employment research.[18] In the U.S., entrepreneurship is today both a science and an industry unto itself.

China's entrepreneurs began popping out of the woodwork almost the second that Deng Xiao Ping uttered his famous phrases in the late '70s that it was "glorious to get rich" and (my favorite), "It doesn't matter whether the cat is black or white. The question is whether it can catch mice!" What is so amazing is not just that entrepreneurs emerged, but that they emerged *at all* after Mao's 30 years of brutal suppression of any expression of individuality, after his failed "Great Leap Forward" in which millions died of starvation, and after the "Cultural Revolution" that gutted a generation of Chinese education, leadership, and culture. Mao's dictatorship came on the heels of more than a century of decline – civil war, warlords, humiliating foreign "concessions," etc. – all deeply unfriendly to any entrepreneurial aspirations, to anyone who wanted to stick his head above the crowd.

Perhaps even more amazing, these entrepreneurs emerged with virtually no formal financing mechanisms available to them. Even today credit is extremely tight and first priority still goes to the dying state-owned sector. It is certainly true than many companies that are now wholly private and very successful began with significant government support, or as state-owned companies. Haier is the best-known example.

Many smaller, successful enterprises never received a penny. Moreover, the venture capital industry is just at the very beginning, with less money being invested in all of China in a year than is invested in Silicon Valley alone in a single month.

In any case, many strong, profitable and totally private companies have grown up fast - like Huwaei, the telecom giant with some 50% of its multi-billion dollar revenue already from foreign sales, and Broad Air Conditioning, the Tasly Group, the Chint Group, the Delixi Group, the Fortune Group, and so on.

What is equally interesting is that even *small, new* companies are showing their confidence and independence. A few years ago, the words "joint venture" were on the lips of every Chinese company executive. You bring the technology and management know-how, we'll throw in the workshop, cheap labor and market access. Today, this kind of traditional combination is rarer and rarer. In a recent partner search for a major multi-national company, we identified and interviewed some 100 small companies, almost all private, to see if they would be interested in a partnership with our prestigious client. There were almost no takers, and we were struck by how dramatically different this was from just a few years earlier. Instead, they said that they had a certain market share and technology expertise that they wished to build on their own. If our client wanted to buy from them, fine. They would welcome this type of "partnership." They had no interest, however, in being "gobbled up" (their words) by a big multinational and losing identity.

Just as in the U.S., most of these entrepreneurial companies are likely to fail. At the same time, some are also very likely to succeed, and to grow and make their mark on the national and international business landscape. In fact, it is very likely that many of the companies that will change the world in 10 to 15 years are unknown or non-existent today.

Beijing has become intensely aware of the value of "home-grown" innovation and job creation. In recent years, aside from the cumbersome banking system, China has moved aggressively to support entrepreneurial development. We mentioned the Shanghai IC design incubator earlier, but incubators are ubiquitous in China. Moreover, in every coastal city they have created and financed special "international incubators" designed especially to attract back to China those citizens who have gone abroad to study and want to return to the homeland to make their fortune. Walk down the halls of these incubators and meet the young Chinese CEOs who graduated from MIT and Stanford and the University of Pennsylvania and …

Entrepreneurs are also increasingly welcomed, and even recruited, into the political arena. Although the numbers are still relatively small, well over 100 private entrepreneurs have become deputies of the National People's Congress (NPC), the nation's top legislative body,[19] and the National Committee of Chinese People's Political Consultative Conference (CPPCC).

To this author, this meteoric rise of entrepreneurship remains a source of considerable mystery. How, why, did this entrepreneurial instinct spring up with such remarkable force through the first chink in sociopolitical policy? Why was it not so suppressed and dispirited that it took several generations to regain any momentum? Why did it not more closely trace Japan after the Second World War, where true entrepreneurs are such a rarity that major government resources must be marshaled to revive any entrepreneurial sense of risk and reward? In China, the new entrepreneurs emerged *prior* to formal government support or financing rather than after it.

If ever there were a challenge to America's status as the world's preeminent entrepreneur, China is it. What will it mean – economically, politically, socially – if the center of gravity of innovation really shifts in her direction and away from the U.S.? We have been successful at answering these challenges in the past, but the odds are different this time if for no other reason than the sheer relative size of the innovation pool – population. Chinese Universities have recovered from the Cultural Revolution and are getting better and better. Already they are graduating tens of thousands of engineers each year. U.S. universities are also graduating a large number of Chinese engineers and scientists each year. Many leading Chinese CEOs in China were educated in America.

The generation that has now come of age in China, born in the '70s and '80s, has largely escaped the debilitation suffered by their parents during the Cultural Revolution. In addition, the one-child policy meant that the Cultural Revolution parents would spare no effort to see that their only child would have a better life in the new China than they did under Mao.

We began to see first hand the emergence of an increasingly talented technical and managerial class in the late '90s. U.S. companies that were setting up operations in China began deliberately to recruit *only* domestic Chinese for top managerial positions. It was an abrupt shift. Before that time, only very expensive U.S. or European expatriates were used to manage such operations. Suddenly they wanted Chinese only. In one case, a major chemical client even specified that they did not even want a mainland Chinese who had been educated abroad – only a domestically educated Chinese plant manager would do.

While the rigid educational examination system still unfairly precludes advancement for many highly capable people in China, the net effect of the system in such a vast population in absolute numbers is still huge. For example, although estimates vary, China is now graduating between 200,000 and 300,000 engineers every year, compared to about 60,000 in the U.S. (We might also mention the rising tide of India, which is graduating around half the number of engineers as China, still about double the U.S. rate.) Yet only about 4% of the population has a college education! What will the world look like when the percentage reaches 10%, or 20%? With this rising reservoir of innovation, it is hard to imagine a scenario that does *not* shift to an Eastern focus.

The Rise of Chinese Multinationals

Since about the year 2000 the results of this rapid emergence in Chinese capability, quality, and confidence has begun to be felt on the global stage. Companies like Haier, Huawei, and Zhongxing (ZTE) are already fairly well-known. Haier is the only Chinese company that has really succeeded well in the U.S. market. Huawei, ZTE, and many others are developing and implementing their Western strategies, starting first in Latin America and other regions. Here are a few highlights:

- In just 20 years time, Haier Group has become the second largest manufacturer of white goods in the world
- In 2005, Haier bid for Maytag, an American business icon
- TCL formed a joint venture with Thomson to become the largest TV manufacturer in the world, gaining instant access to U.S. and EC markets.
- TCL entered into a similar alliance with Alcatel to expand its handset business to the EC and to help keep Alcatel in the handset game
- Ningbo Bird, a large Chinese handset maker, teamed up with Siemens for similar purposes
- Lenovo acquired IBM's PC manufacturing business making it the largest PC maker in the world
- A Chinese consortium wishes to acquire a majority position of Huffy, the old U.S. bicycle maker

- Huawei may take over Marconi, the UK telecom equipment company
- Last, of course, the Chinese National Overseas Oil Company (CNOOC) $18 billion bid for Unocal, the U.S. oil company

The list could go on, and will quickly get longer and longer. What many (but not all) of these examples have in common is that the Western partner gains access to low-cost manufacturing, while the Chinese partner gains access to world renowned brand names and Western markets. Unlike Japan, where Japanese companies laboriously built brand image over a period of decades, China is simply buying into the global branding ballgame.

There are also many Chinese companies that are going it alone, and changing their industries globally. China has certain advantages that are difficult for competitors to address:

- *Cost of labor*, already mentioned, is of course key. These costs are likely to remain low in the aggregate for some time. Skilled labor costs are rising, some quickly, but most labor costs will rise very slowly because of the large supply available.

- *Small profit margins* of 5% and even less are acceptable and often normal in China. For Western companies, it is extremely painful to squeeze margins like this, especially if they are public and shareholders are demanding increasing quarterly returns.

- *Fast learning curve*, as discussed above, has led to rapid improvements in technology, management and, therefore, product quality.

Combined, these factors will present formidable competition in some sectors. There are two major effects from this – one domestic and one global.

First, quality Chinese suppliers quickly replace imports for major foreign OEMs in China. The Motorolas, GMs, DuPonts, etc. in China would all rather buy locally than import because it brings down costs and makes them more competitive. Most have formal policies stating as much. In some cases (e.g. Boeing), local content is required by the Chinese government in return for major purchases. (Such 'offset' contract clauses are not Chinese inventions, incidentally. They have been common in the West for decades.) In other cases, it is pure economics. In either case, the net effect is to draw Western suppliers into the market to supply from the inside, or lose the business!

Tomas Koch of McKinsey uses the specialty chemical market as an example: "Chinese producers increasingly supply the multinational companies that are moving plants [into China] and then dropping their traditional suppliers." They switch "almost as soon as an adequate local source becomes available... Since rising imports to meet growing domestic demand will eventually draw Chinese entrepreneurs with low cost bases into the market for just about all chemicals, any decision to serve the country from faraway plants may be risky."[20]

To show the complexity of this issue of local suppliers displacing Western imports, we can take the chemical example a little further. In general, Chinese competitors win in areas of specialty chemicals where market entry is not too sophisticated. Foreign quality is still key at the high end, and price at the commodity end. The big multinationals still have distinct price and position advantages in commodity chemicals because they have already been producing globally for a long time with lean production systems. The cost of entry can be so high in some of these commodity products that the position of the big players may be relatively safe for a time. Interestingly, for example, Chinese chemical imports nearly tripled from 2000 to 2003, from about $31 billion to about $94 billion, keeping China a net chemical importer. Yet at the same time (2000-2003) turnover of domestic Chinese chemical companies grew from $68 billion to 113 billion, mostly from domestic sales, while profitability grew 80% in 2004 to around $600 million. The threat in this arena, therefore, seems to be more for the middle market and smaller Western producers. It is very likely that Chinese companies will increasingly displace them in China, with ripple effects on their global trade. Survival itself for many of these companies may require that they move aggressively to gain position within China and co-opt some portion of the Chinese competition.

The second effect occurs only when these domestic Chinese suppliers begin to compete in international markets. This quickly puts downward pressure on pricing globally. The upshot is that foreign companies are increasingly driven to expand their China "footprint" not so much to compete in China but to defend their global markets. That is, companies used to go to China to get a piece of the domestic action; now they need to be there to defend their existing global markets.

This latter point is critically important and will not go away. It is the idea that some companies must go to China in order to stay alive *at home*, and in their European markets and their other foreign markets. In the worst case, the choice is between setting up production in China and sacrificing some U.S. employment, or going out of business and sacrificing all U.S. employment.

Key Market Entry Issues

With the above as background, we turn to some of the key challenges that foreign companies must face in tackling the Chinese market?

Intellectual Property Protection

Everyone is concerned about IP protection when they think about a move to China, and rightfully so. There is a common saying about China that, "If it can be copied, it will be." This is true, and it is a problem for Chinese companies as well as for foreign companies. Chinese brand name liquors, cigarettes, electronics, etc. all have their pirated counterparts.

To a degree, IP piracy is a function of stage of national development. We saw the same issues in Japan in the 1970s, for example, and today in the developing economies of Latin America.

In general, the more complex the IP – such as highly engineered systems – the less likely piracy. In fact, much of the universal China pirate image has been generated by just a few sectors where copying is relatively simple, the main ones being software and audio and video media. This latter area is extremely difficult to control, no matter how serious the authorities are about enforcement. Factories can be closed or abandoned in minutes, and spring up in some obscure corner on the other side of town the very next day. In fact, the authorities have closed down many dozens of CD production lines and thousands of illegal print shops.

Another area where IP "theft" is rather common is trademarks. We put the word theft in quotes because it is often not quite outright theft. Alert Chinese would-be entrepreneurs often register names and symbols that are not quite exact rip-offs of names or images, but deliberately close and confusing. This list of victims is long: Home Depot, Minolta, Polaroid, Adidas, Whirlpool, Colgate, 3M, Brother, Godiva, and many more. In most cases, the Chinese entrepreneur (if we can call him that) simply registers officially one or more major names and symbols that are similar but not exactly the same as the target company or brand –

"Polaraid" for Polaroid, for example. In some cases, the Chinese entrepreneur has registered the exact name and image because the foreign company failed to do so itself. In these cases, the foreign companies have sometimes found themselves in the deeply ironic position of actually infringing on their own IP, of being sued for use of their own marks, and having to buy back their own name from the "legal" owner in China![21]

There is much more to the story than the CD and software rip-offs. As we mentioned at the beginning of this essay, there are today more than 500,000 foreign-invested companies in China. Many of the multinationals in fact have multiple operations that must protect IP. Table 7 shows the results of the annual survey of U.S. companies in China by the American Chamber of Commerce ("AmCham"). IP protection ranks fourth in terms of the number of companies who said IP protection was a problem for them. Even at fourth, 76% is a large ratio.

At the same time, the survey showed that only 29% of companies interviewed felt things were improving, but for the second year in a row, that 75% of U.S. companies in China were profitable. To some degree, it is an odd mixture of opinion, but true. Companies are making money and glad to be there, but there is not much IP rights optimism.

Table 7: Top Challenges to Doing Business in China

	% negatively impacted	% who believe conditions are improving
Unclear regulations	92%	51%
Bureaucracy	91%	41%
Lack of transparency	87%	55%
Inconsistent interpretation of regulations	87%	37%
Poor IPR protection	76%	29%
Difficulty enforcing contract terms	72%	22%
Corruption	70%	34%

Source: www.amcham-china.org.cn

To take this further, it is telling to note that this 2005 AmCham survey indicated the following:

- In 2004, wholly foreign-owned enterprises (so-called "WFOEs") surpassed joint ventures as the primary vehicle of FDI, and in 2004 WFOEs accounted for some 70% of FDI

- 94% of respondents said they are optimistic or somewhat optimistic about their China business over the next five years.

- 74% said they were going to increase their investment in China.[22]

Clearly, IP rights must be seen in this larger business context. Increasingly, companies are moving away from JVs and setting up companies they own 100%. We think this indicates a rising comfort level in China by foreign companies, but also a belief (correct for the most part) that IP can best be managed in a wholly owned operation. Overall, these data suggest that it is quite possible to run your business and run it well in China, in spite of IP issues.

China does not like the image of being the world's worst pirate. In fact, piracy brings no larger social or political benefits to China – it is not good for WTO compliance; it is not good for diplomacy; and most of all, it is not good for business. In fact, the government has sincerely tried to improve the situation and has taken many concrete steps to do so. According to many IP experts, China today has as good an IP legal infrastructure as any in the West, all put in place in the last ten years. This includes:

Trademark law	Unfair competition law
Product quality law	Copyright law
Patent law	Customs IP regulations
Special courts	Administrative authorities

To support the laws and regulations, Beijing has instituted a nationwide education and training program for government officials – whether or not they are directly involved in the IP system. They want all government employees to understand the issues and to be alert for problems. The government itself is trying to set the example by being a model of respect for IP – for example, by using only legitimately acquired software products.

The problem is not poor legal infrastructure, or bad intentions. It is enforcement. In a country the size of China it is exceedingly difficult to cover all the enforcement bases. Moreover, China's experience in this area is extremely limited, so she has very little precedent to refer to and very little indigenous expertise in the subject. One prime example of this is the court system. While there is indeed a nationwide system of IP courts, there is not much of a system to prepare properly trained judges. As a result one finds the courts run by young men and women in their twenties, with little experiential reference. It is a kind of on-the-job training in an arena where the success or failure of huge investments is at stake.

It is also very important to understand clearly that not all technology losses are due to illegal piracy. In the case of patents, for example, unless you have explicit patent or trademark protection in the target country, it is perfectly legal for anyone to reverse engineer, copy, improve, etc., your products *if you have not registered your patents or have lost the right to do so.*[23] This, in fact, happens everywhere, in Germany, in Japan, in China and, yes, in the United States, and is perfectly legal.

There are two avenues available to fight piracy – legal and operational. Regarding legal recourse, as anywhere, it is better to avoid it if possible. Even though local protectionism is decreasing and foreign companies are increasingly successful, the process is daunting in terms of both time and money. It is therefore very important to put in place internal corporate policies to protect and monitor IP. Frankly, many of the IP issues faced by foreign companies could have been avoided if they had paid a bit more attention to the issue. That is, IP protection should be treated as a normal, day-to-day management affair. The following are some policy considerations that may be appropriate in various circumstances:

- Screen all employees, as far as possible, for ethical standards
- Register all IP with the appropriate authorities
- Register more rights than you need, such as name and mark variations; anticipate piracy attempts
- Record your registered rights with customs so that they are authorized to seize infringing products

- Track container shipments to and from you
- Maintain an IP portfolio and regularly review and update it. Limit access to the IP portfolio
- Retain any documentary evidence of possible piracy; Chinese courts rely almost exclusively on such physical documentation
- Perform due diligence on and monitor your distribution network, suppliers, subcontractors, licensees, etc.
- Perform due diligence on key hires
- Enter into non-disclosure and non-compete agreements with key staff
- Call out all IP in any JV or collaborative contract and the consequences of infringement
- Build ties to local officials – IP and others – who may help you if you need it
- Form alliances with companies with related problems (foreign and Chinese)
- Phase in your venture: import core technology; limit JV to assembly until your comfort level is high and staff committed
- Train staff on importance of IP protection for their own future
- Watch and protect the facility

You may even wish to consider 24 hour surveillance with Internet "nanny cams." The technology is inexpensive and sometimes just the presence of scanning cameras can be enough to discourage would-be pirates.

If we give all these options due consideration, by far the most important protection against piracy is trustworthy people, suggested by the very first bulleted item. It is critical that foreign companies tie up with companies they know well and trust, or that that they hire honorable people to manage their wholly owned operations. Nothing is more important, even if sometimes very difficult. Trusted relationships, properly rewarded, are the best IP protection.

Guanxi

Guanxi (pronounced 'gwaan-she') means, basically, trusted relationships. Anyone who has had even limited dealings with China knows this term. While personal relationships are very important for business worldwide, they tend to be high art in Asia, and in China have often been a matter of survival itself. It is important to understand this key idea and its changing role.

Simply put, in a society without a basis in law, trusted relationships are the only reliable basis for transactions. Under Mao, there was no real legal reference point for the citizens. There was only an ideological reference point and that was entirely determined not by the people but by Mao personally. As a result, people tended to keep their heads down, because they were never sure how they would be treated if they were noticed. If you were noticed, you ran the risk of being judged, but you were never quite sure if the judgment would be good or bad because *there were no fixed rules*. In fact, Mao often reversed himself entirely on what was good and bad, in terms both of individuals and even his own policies (e.g. the "Let a hundred flowers bloom" movement). In this environment, you have to depend on your friends, your real friends – and even that was often hard to determine under Mao. Trusted relationships became a matter of survival itself.

This is partially why China is such a regional and even local country. Trust, guanxi, is defined by close personal ties, and geography often played a limiting role in developing close relationships. It was hard for someone in, say, Tianjin in the North, to transact business with, say, Guangzhou in the South, because they had no common *basis* for the transaction. In the U.S. the law applies equally, with equal recourse, from Maine to California; it is the common basis for transactions next door or across the continent. In China, the basis was personal guanxi. The guanxi was an invisible contract.

As China develops a law-based society, this dependence on guanxi is changing. Even ten years ago it was more or less in full force. After about 2000, it began to decline somewhat in importance, depending on the sector, the nature of the business, and other factors. That is, the nature of the business opportunity began to carry independent weight and was judged more on business merit than on whether all the key guanxi was in place.

This is not a clear picture. It is now a hybrid mixture of business merit and guanxi, and often hard to tell where one ends and the other picks up. You may need business merit to get in the door, but guanxi to close, or vice-versa. In general, however, the balance is shifting to business merit – to the 'value proposition' in common parlance. Put differently, if you do not have a sound value proposition, no amount of guanxi is likely to get you the deal.

Homework

".. companies often do not fully investigate the market situation, don't perform the necessary risk assessment, and fail to get counsel....*Lack of thorough due diligence is the primary cause of financial loss for American businesses* (emphasis added)."[24]

This quote is from Thomas Lee Boam, former U.S. Minister Counselor for Commercial Affairs in Beijing, and exactly reflects our experience in China as well. While many business people are quick to blame China for their failures, in fact most failures derive from poor homework to begin with – poor homework on the market, on partners, or on the competition.

The problem is true of both large and small companies, but is probably more common among the smaller companies who normally do not have large staff planning capabilities. Sino-foreign partnerships are a prime example. Far too often, foreign companies tie-up with the first company that shows an interest in their products, rather than systematically researching the best fit partner – best fit in terms of market position, technical capability and, above all, trustworthiness, an intangible very difficult to determine and measure. In part, unhappy partnership experiences are behind the trend toward WFOEs, discussed earlier. More important here, many of these unhappy partnerships were simply the result of poor homework.[25]

In any case, one needs to develop a reliable understanding of at least the points listed below:

- **The Market**
    - ✓ How big is it, what are the growth trends and what drives the trends and are these drivers likely to continue?
    - ✓ Production and consumption patterns, key players
    - ✓ What are the key points of product differentiation (see next section)
    - ✓ How does distribution and the supply chain work?
    - ✓ Pricing
    - ✓ How should the target customers be categorized and prioritized?
    - ✓ Geographic concentrations
    - ✓ Common promotional methods

- **The Competition**
    - ✓ Profile the key foreign (like yourself) and domestic competitors; compare pricing and quality, and note especially any progress in quality by domestic competition.
    - ✓ Market share, trends
    - ✓ Does the foreign competition add any local value, or have special alliances with local companies?
    - ✓ How do they sell; what are the channels?
    - ✓ Has anyone had IP issues?

- **Candidate Partners**
    - ✓ Are domestic competitors potential production or distribution partners?
    - ✓ What about other critical players in the supply chain?
    - ✓ Evaluate and compare: trustworthiness, ownership, stability, debt, growth, strengths & weaknesses, etc.

- **Government**
    - ✓ Does the government have special regulations, support programs, restrictions, etc?
    - ✓ What are the key research institutes in your field? (They can often be important and influential.)

Differentiation and Pricing

It is common business wisdom that there are basically only two factors that drive competitive advantage – differentiation and price. It should be clear by now that China is no longer a low-tech market, but it is still a low price market. This combination always creates strategic challenges in China.

The days are gone when an American company could move into China with its old technology. In general, the same rules apply in China as apply in the U.S., and number one among them is *differentiation*. If one imagines that China will be a fast and profitable outlet for products with tight margins here, it is a mistake. The products which sell best in China are the same ones which sell best here – products that are the most *differentiated* in the market. That is, the more a product is differentiated by unique or proprietary know-how or technology or performance or brand image, etc., the more likely it will be a strong candidate product to sell in China. Conversely, the less differentiated it is, the more it will be driven by price because local producers will control the market at, normally, prices one-half or less than those of foreign products.

However, in China the rules on differentiation and pricing are not quite as clean as they are in more developed markets. The term 'The China Price' has become famous and refers to the relentless downward pressure on pricing from buyers. We have seen situations where the "price" was less than the Western company's cost of production, making it impossible to capture the business. The Chinese side always holds out the prospect of huge sales downstream if only you can meet the China Price. It is tempting to take the bait, but there are no guarantees that it will work.

There are two key reasons for the China Price. One is simply that it is a gigantic buyers market. Of course, when the global supply of commodities is strained by Chinese consumption, the price is actually driven up, as we well know. However, for many products, the China Price is simply the result of the buyers' ability to play competitors against each other.

The second reason is, as already discussed, that Chinese companies are moving up the value chain very fast but prices are not increasing proportionately. Thus there is a growing need for Western companies to expand their China "footprint" in order to compete both in China and in world markets. If a product is well-differentiated, the company may not have to invest in local added value for a time. Even then, however, local competition must be monitored carefully and the Western company must be able to move quickly when the threat is apparent – or, preferably, before it is apparent. Otherwise, the market for the foreign company in China will rapidly begin to vanish.

Short and Long Term Strategy

As mentioned at the outset of this essay, despite the fantastic growth in China, entering the market is extremely challenging. Companies need real, workable strategies if they hope to succeed in this market.

Figure 10 shows the three basic modes for entering a market. Virtually any method of getting into a market you can think of will fall into one of these three categories or some hybrid. The left hand column lists the modes with some examples. The right hand column lists some characteristics of that mode, and the center line is a kind of time/commitment continuum.

The strategic challenge is to move from short term sales to long term market position (share) with a maximum return and minimum risk.

Everyone wants to sell and we always recommend that companies sell *before* they invest. This will help them become familiar with the country, the market, competition, pricing, etc., so that any investment will be better informed. Yet selling is not the *only* question, and too often businesses behave as though it is.

In fact, if a company thinks that sales is the only question, then it will not succeed in China in the long term, and maybe not even in the short term. The reason should be clear by now – if you are *only* interested in exporting, eventually someone in China will figure out how

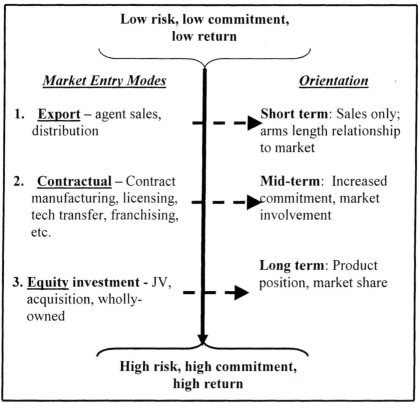

Source: Sino-Consulting, Inc.

Figure 10. Market Entry Modes [iii]

to make what you make, at the same quality level, but at half the price, and your exports will dry up. Therefore, when devising a strategy, a company is wise to consider both short term sales *and* long term product positioning.

Generally speaking, the export or pure sales option always keeps you at arms length from the target market. (In fact, if your primary importer is a stocking distributor, you may know virtually nothing about the actual users of your product, and therefore know learn very little that can help you adapt and grow your export business.) In the short term, it is safe, low risk, and low cost. As you move to contractual and investment modes, you become increasingly involved directly in the target market because you are transferring at least some intellectual assets there. The risk therefore increases, but so does the potential reward.

Summary

The China challenge is a cumulative affair comprising several formidable elements:

- The *'per capita'* **challenge.** This is the impossibility of global per capita parity in certain critical areas, such as fossil fuel consumption, and the vast economic, environmental, social, and political issues this raises. If, for example, China were to consume as much oil on a per capita basis as the U.S., it would require virtually the entire *current daily global consumption*. We deliberately state this not in terms of the 'oil challenge' or the 'steel challenge' because the *per capita* challenge expresses the larger issue. The world, including China, can agree – or *should* agree – that this is neither possible nor desirable. What are the alternatives and how will they play out between China and the West?

- The **entrepreneurial challenge.** The entrepreneurial challenge could as well be called the innovation challenge. America's preeminent position as the world's entrepreneur is almost sure to be challenged by China or, rather, by the millions of educated Chinese driven to find a better life for themselves and their families. In just 20 years, with almost no support infrastructure, private entrepreneurs have already emerged to play an important role in the Chinese economy. What happens if the primary source of technical innovation shifts to the East?

- The **Chinese multinational company challenge**. Much of China's success can be attributed to its policies to support foreign companies. China saw foreign companies not as a threat but as a ticket to the game. Today, at the other end of the spectrum from the entrepreneur, large Chinese companies are beginning to enter the world stage and develop their own brand recognition. The Chinese bid for Unocal seemed like a sudden wake up call for America, especially for the politicians. Partnerships with global leaders are speeding the development of many Chinese players. Unlike the Japanese who did it all on their own, the Chinese are buying into the global marketplace through these partnerships. This is happening far faster than most predicted.

- The **market entry challenge**. Despite the fantastic growth and opportunities, access to and positioning in the Chinese market is very difficult. The buyer's markets, the "China Price," the preference for local suppliers, "guanxi," all make it a tough place to carve out market share.

Postscript

In the mid 1990s I had occasion to host a Japanese business delegation at my home in Philadelphia. Our cordial discussion began to focus more and more on China and her rapid economic growth. The tone of the conversation promptly turned defensive. I asked, "What is your concern about China?" The response from my good friend was: "Its expansionist tendencies." I was stunned. I did not know if he said this because it was the first thing that entered his mind, or because he had been taught this, or because he had come to this conclusion independently, or some combination of these explanations. In fact, his comment was deeply ironic, because it was Japan which had the scarred history of aggression, and there was no historical evidence for such a claim about China.

If the last two centuries are any measure, China was consistently the victim of aggression, not the aggressor. Every major Western power began carving up the imperial carcass in the nineteenth century, with large, independent "concessions" in every major coastal city, followed by Japan's occupation of Manchuria and then invasion south and inland. Chinese power had so deteriorated that she could hardly do anything but accept whatever terms were proposed. There was neither the environment nor the opportunity nor the desire for China to be an aggressor. China just wanted to be left alone.

The fact is, Chinese culture is traditionally inward looking, not externally expansionist. In fact, it was this very trait of looking inward – and feeling culturally superior – that led to the great technological gap between China and the West and the ultimate demise of imperial China. China *refused* to care about what was outside its borders and worried first and foremost about how to keep the outside world out and the inside world quiet! The eminent historian John Fairbank summarized the nineteenth century Qing Dynasty in this respect as follows:

"They had little sympathy with entrepreneurs, kept their own people out of trade, and *penalized anyone who went abroad.* All in all, their influence seems to have been backward- and *inward-looking, defensive, and xenophobic.*"[27] (Italics added.)

Chairman Mao continued this fundamental imperial tradition. He had profound distrust of anything non-Chinese. This was made painfully clear during the Cultural Revolution when anyone who had ever had any contact whatsoever with the outside world was, at least, deeply suspect, and at worst brutalized as a "capitalist roader" or outright spy on purely imaginary evidence.

Mao was once asked why he had never traveled outside of China. He responded to the effect that he could spend his entire life traveling inside China and still not comprehend her, so why go outside? Mao built a powerful army not to conquer but to make sure the country would never again be kicked around as it had been for the previous century.[28]

There are no more profound symbols of this inward-looking character than the Great Wall and the Forbidden City. The former was intended to keep out the "barbarians," the latter to keep out the riffraff, and in both cases so those inside could tend their own cultural gardens as they saw fit.

You can even see this in the global Diaspora of "China towns." Each one imports and re-creates its home environment, often until it looks, feels, and tastes like the homeland. The residents integrate into the larger social context as proper civility requires, and show due respect for the institutions that have allowed them to replicate their culture. At the same time, their essential social and political interests are focused inward to the community, and even more particularly to the family.

In this cultural context, China never developed a coherent expansionist political theory or policy. The primary focus was on guarding borders and fending off aggression, not on aggressive acquisition of new territory. Their interest in contiguous territories was historically viewed in this defensive context, not in terms of conquest. By way of example, although this author would very much like to see an independent Tibet, it must be understood that from China's point of view Tibet is both *part of* China, like Taiwan, and an historical buffer state.

China never developed a policy comparable to, for example, America's nineteenth century doctrine of Manifest Destiny that permitted this country's continental expansion with impunity. China was already thousands of years old when the United States was in its first century, and it hadn't the least interest in what we were doing.

The first point is a rather simple one: Even today, China has no interest in military expansion beyond its borders. Just as 2000 years ago, they have more than they can handle within their own borders. (Taiwan, of course, is considered to be within the natural and historical borders of China.)

This is by no means to suggest that China is all gentle and peace loving but, that as far back as one can trace, its violence has been focused inward against its own people, not against outside enemies. Rebellions in Imperial China were put down with chilling ferocity. For the last 100 years, since the demise of the Imperial system, China has suffered through constant, non-stop, violent upheaval, beginning with the founding of an unstable republic, proceeding through regional warlords, then to the unspeakable atrocities of the Japanese invasion and the loss of over 12 million people (some 10 million of whom were civilians), and on to nationwide civil war between the Kuomintang and the communists, then through Mao's disastrous "Great Leap Forward" in which millions more starved to death, and finally to his "Cultural Revolution" which victimized millions more and deprived the country of an entire generation of educated citizens. Anyone who has read first hand accounts of the Cultural Revolution – like Nien Cheng's *Life and Death in Shanghai* or Jung Chang's *Wild Swans: Three Daughters of China* – gets an idea of just how xenophobic and internally brutal the Chinese can be against their own citizens. In summary, China has shown as much violence and barbarity as any other country, but almost entirely within its own borders. There were no pretensions to any externally focused ideologies, no "Greater Asian Co-prosperity Spheres," for example, in China's politics as there was in Japan's.

The second point is that we ought to view the human rights issue in this context of two centuries of civil conflict and foreign invasion. If we think back just one generation, the China under Mao was a human rights nightmare. The difference now, in just one generation, is like night and day. We should congratulate China on its fast progress and encourage her to continue to do more, instead of slapping her wrists for not going further and going faster. In fact, if we compared how long the West took to achieve a comparable level of human rights, we would not fair well at all.

Under an increasing rule of law, it seems that China is at last coming into its own. After seemingly endless chaos as Imperial rule slowly disintegrated and new systems struggled to take its place and unify the country, China has finally been able to create an environment that allows its hard working people to dream once again. And China's dreams, we can be sure, will be the biggest challenge of all.

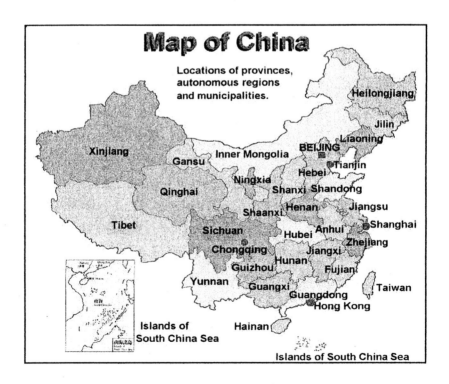

References

1. http://en.wikipedia.org/wiki/List_of_countries_by_GDP(nominal)
2. International Monetary Fund, as reported in *China: Is the World Really Prepared?*, Amy Raskin and Brad Lindenbaum, Bernstein Investment Research and Management, 2004, p 2.
3. *Millennium Development Goals, China's Progress,* United Nations Country Team in China; Office of the United Nations Resident Coordinator, Beijing, 2004; preface, p16, et al. The report further noted that "...between 1978 and 2000 the number of poor fell from 250 million to 30 million. The World bank raised China's classification to a lower middle-income country in 1999, when income per capita surpassed the U.S. $755 cut-off point for low-income countries." (Or see www.undp.org/mdg/chinaMDG.pdf)
4. These ratios are from various sources and different timeframes, but they don't need to be current to illustrate the point well. The first four items are from the *Far Eastern Economic Review*, 2002; the hard disk drive percentage is also a 2002 (November) number from a *BBC News* article at http://news.co.uk/1/hi/business/2415241.stm. The cell phone percentage is for 2004 and is from *isupply* estimates (www.isupply.com/marketwatch/default.asp?id=204), although similar numbers can be found in many places. The digital camera numbers are from the *Peoples Daily,* English version, December 22, 2004, or http://English.people.com.cn. The office equipment percentage is from Forbes 2000, 04.05.04, Mitchell Martin, *The World Tilts Toward China.*
5. http://english.ccidnet.com
6. www.icc.sh.cn
7. New York Times Magazine, April 3, 2005, p 37
8. This report is published every year by the Electronics Information Product Management Department of the Ministry of Information Industry (MII) and China Software Industry Association (CSIA).
9. See especially No. 18 Notice to encourage the development of the software industry with preferential policies, issued by the State Council in 2000, and No. 47 File to promote the development of the software industry with specific measures, issued by the Information Office of the State Council in 2002. See also the Government Procurement Law officially put into effect on January 1, 2003, which gives clear preferential treatment to domestic companies. The key agencies involved are the Ministry of Information Industry (MII) and the Ministry of Science & Technology (MOST).

10. CCW Research is an information center under China Computer World network (CCW www.ccw.com.cn/). CCW is founded by China Computer World Publishing and Services Co. In brief, CCW publishes and sells information about hardware, software, training, services.
11. UFSoft is the largest supplier in China of management software, ERP software and financial software. The enterprise application software spectrum of UFSoft is very wide, including ERP (Enterprise Resource Planning), SCM (Supply Chain Management), CRM (Customers Relations Management), HR (Human Resources), EAM (Enterprise Asset Management), OA (Office Automation), industry management software and so on.
12. This was a 2004 IDC survey conducted under contract to the American Business Software Alliance (BSA).
13. Projections from Electric Power Technology Market Association of China. Stated from a different angle, the Association points out that some 2/3 of new capacity which will be built between 2000 and 2020 will be built *after* 2005.
14. Diamond, Jared, *Collapse*, Viking Press, 2005. See chapter entitled "China: Lurching Giant."
15. Credit Suisse / First Boston, *China's Capacity Expansion*, May 2005, p 67.
16. Lester Brown founder & president of EPI, as quoted by Abid Asiam, *OneWorld US, Thursday, March 10, 2005.*
17. An excellent discussion of the definition complexities of enterprises in China, and somewhat more conservative estimates than are shown in Table 5, can be found in OECD's 2005 Economic Survey of China, chapter 2. See especially Annex 2.A1, p 125.
18. The author was honored to be a small part of the founding of the ground breaking National Business Incubator Association (NBIA) in the mid 1980s, in support of Dr. Randall Whaley who became the NBIA founding Chairman. Dr. Whaley was CEO of the University City Science Center (UCSC) in Philadelphia. UCSC was the first urban research park in America and, many say, the first business incubator. Hence Dr. Whaley's invitation to be the founding chair.
19. The NPC has traditionally been a rubber stamp for the State Council and the Party, but in recent years has been gaining more and more respect and real power.
20. Tomas Koch, *McKinsey Quarterly: China Today,* 2004 Special Edition, pp90, 93.

21. These examples come from a presentation of the IP attorney Douglass Clark, with Lovells in Shanghai. The author and Mr. Clark have on a number of occasions been part of the faculty of China business seminars sponsored by the Structured Finance Institute.
22. http://www.uschina.org/public/documents/2005/08/2005_membersurvey.pdf
23. Normally, patent holders have one year from the date of issuance to register their patents in most foreign countries, including China.
24. This is quoted from a power point presentation given by Mr. Thomas Lee Boam, former Minister Counselor for Commercial Affairs in the U.S. embassy in Beijing, who was discussing the challenges of doing business in China.
25. Our favorite example is what we like to call the "Uncle Bill" syndrome. A CEO decides to go to China but doesn't have any idea how to go about it. The he remembers that his 'Uncle Bill' once traveled to Beijing and met a guy who knew a guy who knew a guy, and one of the guys is inevitably a "high government official," (often a relative of Hu Jintao). So the CEO decides, "I'll give Uncle Bill a call..." and so on. (When we describe this to corporate groups, heads around the table invariably start the dashboard nod, "Oh, yes!") Networks and networking are of course indispensable. But it is just a one piece of the puzzle.
26. SCI developed this continuum based on the wonderful work of Professor Franklin Root, Entry Strategies for International Markets, 1994, Jossey-Bass, Inc., San Francisco.
27. John King Fairbank, *The Great Chinese Revolution: 1800-1985*, Harper & Row, NY, 1986, p 36.
28. The decision to enter the Korean Conflict on the side of North Korea is said to have been one of the most difficult decisions Mao ever made. He knew the consequences could be global war, even nuclear war, and did not want to do it. It is one of the few examples we have of any military movement outside China's borders, other than the brief "punitive" assault on Vietnam in 1979.

Indexes

Author Index

Jones, Roger F., 1
Kermani, Faiz, 47

Weckesser, Timothy C., 111
Wollowitz, Susan, 47

Subject Index

A

Acquire-and-divest strategies, chemical companies, 31–34
Africa, economic growth, 27
Agency for Science, Technology and Research (A*STAR), Singapore, 95
Aging, global population projections, 55f
AIDS, funding research activities, 102
Air conditioners, China's estimated percentage of global production, 114t
Alaska's Arctic National Wildlife Refuge (ANWR), exploration and development, 14
American Chemical Council, Responsible Care program, 23–24
American Chemical Society (ACS)
 employment survey, 18, 20
 salary survey, 20
Arizona
 attempting to create biotech hotbed, 71
 new Medical School in Phoenix, 72
Asian countries
 emerging biotech companies, 92–93
 employment in biotech industry, 61f
 growth opportunities, 26–27
 interest to pharmaceutical companies, 57
 stem cell research, 100
Association of British Pharmaceutical Industry (ABPI), United Kingdom (UK), 80, 82
Australia, employment in biotech industry, 61f

Automobiles
 China, 115–117
 market share in China, 116f

B

Baby Boomer generation, retirement of, 21
BASF
 comparison with DuPont and Monsanto, 32–34
 nylon fibers and Honeywell, 33
 polyolefins, 32–33
 stock price performance, 35f
 verbund (integration) concept, 32
Big picture, China, 112–113
BIOCOM, Southern California biotech community, 68
BioIndustry Association (BIA), United Kingdom (UK), 81
Biomedical Research Council (BMRC), Singapore, 95–96
Biopharma, biological and pharmaceuticals, 39–40
Biopolis
 creation of, in Singapore, 54
 Singapore, 95
Bioscience Innovation and Growth Team (BIGT), United Kingdom (UK), 81, 82t
Biotech, term, 50
Biotech companies
 California as biotech state, 67–68
 clinical development profile for, in Europe, 87f
 employment in biotech industry, 61f
 financial maturation, 73

relationship between National Institute of Health (NIH) funding and, 66, 67*t*
research and development investment, 50*t*
venture capitalists and start-up enterprises, 54
Boston, biotech hotbed, 68
Byrd Amendment, tariffs as result of, 7

C

California
  biotech state, 67–68
  employment in U.S. pharma industry, 65*f*
  relationship between NIH funding and biotech companies, 66, 67*t*
  research and development "hotbed", 54, 66
  stem cell research, 100
Cameras, China's estimated percentage of global production, 114*t*
Canada
  2004 top 10 GDP countries, 113*t*
  employment in biotech industry, 61*f*
Candidate partners, developing understanding, 151
Capital
  chemical industry, 27–28
  Massachusetts' biotech businesses, 69
  privately owned companies raising, 30–31
  U.S. private investor contribution stimulating R&D, 72–73
Cell phones, China as Motorola's global production base, 115
Census Bureau data, U.S. manufacturing, 3, 4*f*

Centre National de Recherche Scientifique (CNRS), France's research centre, 78–79
Cheap revolution, global competition, 23
Chemical companies, acquire-and-divest strategies, 31–34
Chemical industry
  crude oil and natural gas prices, 11–12
  duplicative or differing visions, 31–34
  employment by firm size, 19*f*
  employment trends, 18, 20
  environment, 23–25
  escalation of competition, 22–23
  future objectives, 38–41
  globalization and regional markets, 25–27
  global strategy, 38–39
  identifying business sectors for globalized market, 39
  influence of financial community, 27–31
  joint ventures, 40–41
  manufacturing changes, 2
  merger and acquisition (M&A) program, 40
  National Association of Manufacturers (NAM), 6
  outsourcing R&D, 21
  political reforms, 41
  professional employment in U.S., 42
  quality improvement and cost reduction, 39
  regional, R&D spending totals, 36*f*
  specialty markets, 39–40
  U.S., vs. overall U.S. manufacturing, 11–12, 13*f*
  U.S. chemical R&D, 34, 37–38, 43
Chemistry graduates, salary survey, 20
China
  2004 top 10 GDP countries, 113*t*
  currency valuation policies, 6–7

DuPont and GE, 37
efforts to acquire U.S. firms, 8
estimated percentage of global production of selected products, 114*t*
fierce competition, 25–26
growing consumer class, 25
growth of manufacturing sector, 2–3
interest to pharmaceutical companies, 57
investment of trade surplus dollars, 8
labor and manufacturing costs, 6
map, 158
per capita energy consumption, 125, 126*f*
pharmaceutical industry, 96–98
pharmaceutical research and development (R&D), 98
polyolefin capacity, 15
science universities, 21
top challenges of doing business in, 145*t*
United States Patent and Trademark Office (USPTO), 97
U.S. imports, 9, 10*f*
U.S. manufacturing and, 6–11
China challenge
automobiles, 115–117
automotive market share, 116*f*
big picture, 112–113
candidate partners, 151
Chinese multinational company challenge, 155
competition, 151
differentiation and pricing, 151–152
economic trends and implications, 112–113
entrepreneurial challenge, 138–141, 154–155
environmental challenge, 131–134
estimated percentage of global production of selected products, 114*t*
fixed and mobile phone growth in China, 118*f*
government, 151
guanxi, 148–149
handset market shares, 119*f*
homework, 150–151
intellectual property protection, 144–148
market, 150
market entry challenge, 155
market entry modes, 153*f*
'*per capita*' challenge, 154
postscript, 155–158
power and pollution, 124–131
rise of Chinese multinationals, 141–144
rise of public sector, 134–137
short and long term strategy, 152–154
software industry, 120–123
telecommunications, 117–120
trusted relationships, 148–149
world's workshop, 113–115
See also Power and pollution
China Price, competitive nature of globalization, 22
Chinese multinationals, rise of, 141–144
Coal
dependence of China, 129, 131
five-year plan for energy efficiency in China, 131
mining safety problems in China, 128–129
Coal gasification, costs, 12, 14
Collapse, Jared Diamond on China, 132
Committee on Science, Engineering and Public Policy, recommendations, 63, 64*t*

Commodities, China's estimated percentage of global production, 114*t*
competition
  China, 25–26
  developing understanding, 151
  escalation of, 22–23
  pharmaceutical and biotech sectors, 53–54
Connecticut
  employment in U.S. pharma industry, 65*f*
  relationship between NIH funding and biotech companies, 66, 67*t*
Contractual market entry mode, China, 153*f*
Controlled float, yuan against dollar, 9
Corruption, China, 26
Cost containment
  healthcare in Europe vs. United States, 63–64
  public health and benefits collide, 55
Cost of labor, advantages of Chinese companies, 142
Cost pressures, pharmaceutical industry, 76
Crude oil prices, chemical industry, 11–12
Currency valuation
  Chinese government, 6–7
  Japan, 7
Current Employment Statistics (CES), U.S. Bureau of Labor Analysis (BLA), 17–18
Current Population Survey (CPS), U.S. Bureau of Labor Analysis (BLA), 17–18

D

Denmark, clinical development profile for biotech companies, 87*f*
Department of Defense, pharmaceutical industry interactions, 61
Diamond, Jared, *Collapse*, 132
Differentiation, business in China, 151–152
Digital cameras, China's estimated percentage of global production, 114*t*
Diseases, early intervention, 52–53
Doctorates, National Science Foundation (NSF) Survey of Earned, 62–63
Dow Chemical, intrapreneuring, 43
Dow Jones Industrial Average (DJIA)
  DuPont, Monsanto, BASF, 35*f*
  Monsanto and DuPont common shares, 33–34
Drilling restrictions, oil and gas, 14–15
DuPont
  comparison with Monsanto and BASF, 32–34
  Dow Jones Industrial Average (DJIA) for, and Monsanto, 33–34
  stock price performance, 35*f*
  technical center expansion in Asia, 37

E

Early intervention, disease treatment, 52–53
Earth Policy Institute (EPI), extrapolations for China, 133–134
Eastern European countries, future growth potential, 27
Economic growth rates, China and India, 25
Economic trends and implications, China, 112–113
Economy, service sector vs. manufacturing, 3

Employment
  biotech industry, 61f
  overall and chemistry professionals, 16–22
  pharmaceutical, in Europe, 74f
  surveys by U.S. Bureau of Labor Analysis (BLA), 17–18
  uncertain outlook, 20
  U.S. pharmaceutical sector, 63
  U.S. pharma industry, 65f
Entrepreneurial challenge, China, 138–141, 154–155
Entrepreneurial start-up, pharmaceutical R&D, 54
Environment
  chemical industry and, 23–25
  China and *Collapse* by Jared Diamond, 132
  China challenge, 131–134
  fossil fuel paradigm, 133
Equity investment, market entry mode in China, 153f
Ethical drugs, under patent, 50–51
Europe
  boosting pharmaceutical innovation in, 76–80
  clinical development profile for biotech companies, 87f
  complexity of pharmaceutical industry, 73–76
  employment in biotech industry, 61f
  financial comparisons for biotech sectors in U.S. and, 75t
  G10 Medicines Group recommendations for pharmaceutical competitiveness, 77, 78t
  global population projections, 55f
  healthcare cost containment, 63–64
  key market for international pharmaceutical industry, 58–59
  pharmaceutical expenditures for R&D, 60f
  pharmaceutical R&D spending, 57f
  price controls and pharmaceutical industry, 59
European Commission (EC), Framework Program, 79
European Federal of Pharmaceutical Industries, research and development in Spain, 88
European Federation of Pharmaceutical Industry Associations, German pharmaceutical industry, 87
European Union (EU)
  pharmaceutical industry, 73–76
  REACH policy (registration, evaluation, and authorization of chemicals), 24–25
Exploration restrictions, natural gas, 14–15
Export, market entry mode in China, 153f

F

Farmaindustria, pharmaceutical industry in Spain, 88
Federal government policies, National Association of Manufacturers (NAM), 5–6
Financial analysts
  chemical industry, 28–29
  short-term results, 28
Financial community, influence on chemical industry, 27–31
Firms
  chemical industry employment by firm size, 19f
  U.S. chemical industry vs. overall U.S. manufacturing, 11, 13f
Fixed phone, growth in China, 118f
Florida
  attempting to create biotech hotbed, 71
  relationship between NIH funding and biotech companies, 66, 67t

Foreign direct investment (FDI), China, 112
Foreign invested companies (FIEs)
  China, 135–136
  GDP contribution in China, 135t
Fossil fuel paradigm, environment, 133
France
  2004 top 10 GDP countries, 113t
  clinical development profile for biotech companies, 87f
  environment for research and development (R&D), 83–86
  healthcare spending, 83
  LEEM (Les entreprises du médicament), 83, 84
  pharmaceutical employment and R&D expenditure, 74f
  pharmaceutical production, 84f
  pharmaceutical R&D spending, 57f
  PharmaFrance recommendations, 85t
Friedman, Thomas, *The World is Flat*, 120
Future Shock, Alvin Toffler, 2

G

G10 Medicines Group, recommendations for improving European pharmaceutical competitiveness, 77, 78t
Gas turbine production, China, 129, 130f
GE, research and development staff, 37
Genetically modified organisms (GMOs), Europe and U.S., 75–76
Georgia, relationship between NIH funding and biotech companies, 66, 67t
German Association of Research-based Pharmaceutical Companies, VFA, 86–87

Germany
  2004 top 10 GDP countries, 113t
  chemical industry R&D, 38
  clinical development profile for biotech companies, 87f
  environment for research and development (R&D), 86–87
  European Federation of Pharmaceutical Industry Associations (EFPIA), 87
  German Association of Research-based Pharmaceutical Companies (VFA), 86–87
  manufacturing, 5
  new jobs at GE, 37
  pharmaceutical employment and R&D expenditure, 74f
  pharmaceutical production, 84f
  pharmaceutical R&D spending, 57f
Global competition, stem cell research, 99–101
Globalization
  chemical industry challenge, 42–43
  competitive nature of, 22–23
  effect of, on pharmaceutical R&D activity, 56–57
  regional markets and, 25–27
Global projections, population, 55f
Global strategy, chemical industry, 38–39
Government policies
  cost burdens, 43
  developing understanding, 151
  industry competitiveness, 41–42
Government support, software industry in China, 122–123
Graduates, chemistry, salary survey, 20
Greenspan, Alan, revaluation of yuan, 8
Guanxi, China, 148–149
Gulf Coast oil and gas fields, Hurricanes Katrina and Rita, 15–16

H

Handset, market shares, 119f
Hard disk drives, China's estimated percentage of global production, 114t
Healthcare, benefiting from innovation, 52–53
Homework, developing understanding of China, 150–151
Honeywell, nylon fibers and BASF, 33
Huntsman Chemical, privately held companies, 30–31
Hurricane Katrina, Gulf Coast oil and gas fields, 15–16
Hurricane Rita, Gulf Coast oil and gas fields, 15–16

I

Illinois
  employment in U.S. pharma industry, 65f
  home to pharmaceutical company, 66
India
  2004 top 10 GDP countries, 113t
  DuPont and GE, 37
  growing consumer class, 25
  interest to pharmaceutical companies, 57
  pharmaceutical industry, 93–94
India Bhopal accident, Union Carbide, 24
Indiana, employment in U.S. pharma industry, 65f
Innovation
  boosting pharmaceutical, in Europe, 76–80
  constraints on, 55
  cost containment, 55
  intellectual property rights, 56
  life blood of pharmaceutical industry, 48–54
  process of research and development (R&D), 51–52
  public health benefits from, 52–53
  regions wanting pharmaceutical R&D, 53–54
  states creating, centers, 70–72
  stem cell research, 99–101
  venture capitalists and start-up enterprises, 54
Integrated circuits, China, 120
Integration concept, BASF, 32
Intellectual property protection
  China, 144–148
  fighting piracy, 147–148
  innovation and profit, 56
  piracy in China, 146–147
  top challenges of doing business in, 145t
International AIDS Vaccine Initiative (IAVI), funding, 102
Intrapreneuring, Dow Chemical, 43
Investment
  alternate models of, 101–103
  pharmaceutical research and development, 103–104
Investment policies, disfavoring U.S. chemical companies, 29
Investors, chemical industry, 28–29
Ireland, pharmaceutical production, 84f
Italy
  2004 top 10 GDP countries, 113t
  clinical development profile for biotech companies, 87f
  pharmaceutical employment and research and development (R&D) expenditure, 74f
  pharmaceutical production, 84f
  pharmaceutical R&D spending, 57f

J

Japan
  2004 top 10 GDP countries, 113*t*
  chemical industry research and development (R&D), 38
  chemical industry R&D spending totals, 36*f*
  currency valuation, 7
  economic invasion, 9
  global population projections, 55*f*
  Japan BioVenture Development Association (JBDA), 91–92
  key market for international pharmaceutical industry, 58–59
  manufacturing, 5
  per capita energy consumption, 125, 126*f*
  pharmaceutical expenditures for R&D, 60*f*
  pharmaceutical industry, 90–92
  pharmaceutical R&D spending, 57*f*
  R&D, 90–91
Japan BioVenture Development Association (JBDA), collaboration, 91–92
Jobs
  chemical industry employment by firm size, 19*f*
  U.S. chemical industry vs. overall U.S. manufacturing, 11, 13*f*
Joint ventures, chemical industry, 40–41

L

Learning curve, advantages of Chinese companies, 142
Les entreprises du médicament (LEEM), French pharmaceutical industry, 83, 84
Liability costs, U.S. manufacturing, 5–6
Life expectancy, innovation in healthcare, 52
Liquefied natural gas (LNG), importation, 14
Long term strategy, business in China, 152–154

M

Management
  identifying business sectors for globalized market, 39
  overdoing movement to broaden experience, 30
Managers, "fast track", 30
Manufacturing
  advances in productivity, 2
  changes in U.S., 4*f*
  chemical industry outperforming overall, 11–12, 13*f*
  China and U.S., 2–3
  Germany, 5
  Japan, 5
  outsourcing jobs, 2
  service sector growth vs., 3
  United States base, 2–6
  U.S., and China, 6–11
  U.S. Census Bureau data, 3, 4*f*
  workers' income, 5
Map, China, 158
Market, developing understanding, 150
Market entry challenge, China, 155
Market entry modes, China, 153*f*
Maryland, relationship between NIH funding and biotech companies, 66, 67*t*
Massachusetts
  employment in U.S. pharma industry, 65*f*
  "Genetown", 68
  key challenges for, 70*t*
  pharmaceutical industry, 68–70

relationship between NIH funding and biotech companies, 66, 67t
research and development "hotbed", 54, 66, 68–70
status report on biotechnology industry, 69, 70t
Medical school, new in Phoenix, Arizona, 72
Merger and acquisition (M&A) program, chemical industry, 40
Mexico, economic development barriers, 27
Michigan
attempting to create biotech hotbed, 71
pharma R&D environment, 71–72
Middle East producers
olefin manufacturing, 14–15
U.S. chemical industry vs., of olefin products, 42
Mining, coal, safety problems in China, 128–129
Ministry of Health, Labor and Welfare (MHLW), Japanese research and development, 92
Ministry of Information Industry (MII), China's software industry, 122–123
Mobile phones
China's estimated percentage of global production, 114t
growth in China, 118f
Models, investment, 101–103
Monsanto
comparison with DuPont and BASF, 32–34
Dow Jones Industrial Average (DJIA) for, and Dupont, 33–34
stock price performance, 35f
Multinationals
Chinese multinational company challenge, 155
rise of Chinese, 141–144

N

National Association of Manufacturers (NAM), U.S. federal government policies, 5–6
National Institute of Diabetes and Digestive and Kidney Disease, Phoenix, Arizona, 72
National Institutes of Health (NIH)
basic research funding, 62
pharmaceutical industry interactions, 61
relationship between NIH funding and biotech companies in states, 66, 67t
National Science Foundation (NSF)
NSF Survey of Earned Doctorates, 62–63
pharmaceutical industry interactions, 61
reductions in funding, 62
Natural gas
chemical industry and, prices, 11–12
chemical uses, 14
exploration and development, 14
liquefied (LNG), 14
projected consumption in China, 129, 130f
New Jersey
employment in U.S. pharma industry, 65f
relationship between NIH funding and biotech companies, 66, 67t
New York
employment in U.S. pharma industry, 65f
relationship between NIH funding and biotech companies, 66, 67t
North America, global population projections, 55f
North American Free Trade Association (NAFTA), Mexico and U.S., 27

North Carolina
  employment in U.S. pharma industry, 65*f*
  relationship between NIH funding and biotech companies, 66, 67*t*
  research and development "hotbed", 66
  Research Triangle Park (RTP), 71
Nylon fibers, BASF and Honeywell, 33

O

Office equipment, China's estimated percentage of global production, 114*t*
Olefin manufacturing
  European producers, 16
  Middle East competition, 14–15
  polypropylene (PP) and polyethylene (PE) producers in U.S., 15
  U.S. chemical industry vs. Middle East, 42
Organization of Economic Cooperation for Development (OECD)
  health care, 76
  healthcare spending, 83
  innovation policy and economic performance, 77
  life expectancy, 52
Outsourcing, manufacturing jobs, 2

P

Pacific countries, employment in biotech industry, 61*f*
Pacific Rim, U.S. imports, 9, 10*f*
Patent life, pharmaceuticals, 50–51
Payrolls, U.S. chemical industry vs. overall U.S. manufacturing, 12, 13*f*

Pennsylvania
  employment in U.S. pharma industry, 65*f*
  home to pharmaceutical company, 66
  relationship between NIH funding and biotech companies, 66, 67*t*
Per capita challenge, China challenge, 154
Perception, chemical industry, 42
Pharmaceutical companies
  global investment in research and development (R&D), 48–49
  research and development investment for top global, 49*t*
Pharmaceutical industry
  alternate models of investment, 101–103
  boosting innovation in Europe, 76–80
  California, 67–68
  China, 96–98
  cost containment, 55
  effect of globalization on research and development (R&D) activity, 56–57
  emerging Asia, 92–93
  environment for R&D in Germany, 86–87
  environment for R&D in France, 83–86
  environment for R&D in Spain, 88–90
  European complexity, 73–76
  India investing for future, 93–94
  innovation, 48–54
  intellectual property rights, 56
  Japan, 90–92
  Massachusetts, 68–70
  mounting cost pressures, 76
  public health benefits from innovation, 52–53
  R&D investment for top biotech companies, 50*t*

R&D investment for top global pharmaceutical companies, 49*t*
recommendations for improving European, competitiveness, 77, 78*t*
regions wanting R&D, 53–54
Singapore, 95–96
state investment in U.S., 65–67
states creating innovation centers, 70–72
stem cell research, 99–101
trans-oceanic dynamics, 58–59, 60*f*
U.S. private investor contribution stimulating R&D, 72–73
U.S. R&D dominance, 60–64
venture capitalists and start–up enterprises, 54
See also Innovation
Pharmaceutical Industry Competitiveness Taskforce (PICTF), United Kingdom (UK), 80, 83
Pharmaceutical production, Europe in 2003, 84*f*
PharmaFrance, recommendations for French biopharmaceutical sector, 85*t*
Phoenix, Arizona, pharma R&D activities for local economy, 72
Political reforms, chemical industry, 41
Pollution
China, 26, 132–133
See also Power and pollution
Polyethylene (PE), U.S. producers, 15
Polyolefins, BASF, 32–33
Polypropylene (PP), U.S. producers, 15
Population, global projections, 55*f*
Portfolio management, pharmaceutical companies and management firms, 51–52
Power and pollution
China, 124–131
coal consumption, 128

dependence of China on coal, 129, 131
energy sources in China, 127*f*
installed generation capacity in China, 124*f*
mine safety, 128–129
per capita energy consumption, 125, 126*f*
projected installed capacity of gas turbine power plants in China, 129, 130*f*
projected natural gas consumption, 129, 130*f*
projected sources of energy in China, 127*t*
safety problems in coal mining, 128–129
simulated growth in total energy consumption, 128*f*
See also China challenge
Press reports, moving U.S. manufacturing offshore, 41
Price controls, pharmaceutical industry's view of Europe, 59
Pricing, business in China, 151–152
Private investor, U.S., contribution stimulating R&D, 72–73
Privately held companies
employment distribution in China, 137*t*
GDP contribution in China, 135*t*
Huntsman Chemical, 30–31
raising capital, 31
Private sector, rise of, in China, 134–137
Productivity, manufacturing advances, 2–3
Profarma program
goals for innovation in Spanish pharmaceutical sector, 90*t*
Spain, 89
Professional employment, U.S. chemical industry, 42
Profit margins, advantages of Chinese companies, 142

Public health
  benefiting from innovation, 52–53
  cost containment, 55
Publicly held companies
  acquire-and-divest strategies, 31–34
  employment distribution in China, 137*t*
  GDP contribution in China, 135*t*
  investor interest, 42
  quarterly challenge, 29–30
Public policy, recommendations for changes, 63, 64*t*

Q

Quality improvements, chemical industry, 39

R

REACH policy (registration, evaluation, and authorization of chemicals), European Union (EU), 24–25
Refineries, Hurricanes Katrina and Rita, 15–16
Refrigerators, China's estimated percentage of global production, 114*t*
Regional markets, globalization and, 25–27
Report from Committee on Science, Engineering and Public Policy, recommendations, 63, 64*t*
Reputation, chemical industry, 42
Research and development (R&D)
  chemical industry in United Kingdom, 38
  Chinese pharmaceutical R&D, 98
  concepts of portfolio management, 51–52
  effect of globalization on pharmaceutical R&D activity, 56–57
  environment for R&D in France, 83–86
  environment for R&D in Germany, 86–87
  environment for R&D in Spain, 88–90
  environment for R&D in United Kingdom, 80–83
  Germany chemical companies, 38
  Indian pharmaceutical industry, 94
  investment for top biotech companies, 50*t*
  investment for top global pharmaceutical companies, 49*t*
  Japanese chemical companies, 38
  Japan pharmaceutical industry, 90–92
  pharmaceutical expenditures for, in U.S., Europe, and Japan, 60*f*
  pharmaceutical R&D expenditures in Europe, 74*f*
  pharmaceutical spending in key global markets, 57*f*
  regional chemical industry spending totals, 36*f*
  United States R&D dominance, 60–64
  U.S. chemical industry, 34, 37–38, 43
  U.S. private investor contribution stimulating, 72–73
Research Triangle Park (RTP), North Carolina, 71
Responsible Care program, American Chemical Council, 23–24
Russia, 2004 top 10 GDP countries, 113*t*

S

Safety, coal mining problems in China, 128–129
Salary survey, American Chemical Society (ACS), 20
Sales, U.S. chemical industry vs. overall U.S. manufacturing, 12, 13*f*
San Diego, biotech hotbed, 68
San Francisco bay area, biotech hotbed, 68
Securities and Exchange Commission (SEC)
  disfavoring U.S. chemical companies, 29
  individual and institutional investors, 28
Service sector of economy, manufacturing vs., 3
Short term strategy, business in China, 152–154
Singapore
  Biopolis, 95
  creation of Biopolis, 54
  pharmaceutical industry, 95–96
Smoot–Hawley tariffs, global trade war, 7
Software industry
  China, 120–123
  government's support measures, 122–123
  value in China, 121*t*
South America, economic growth, 27
South Korea
  per capita energy consumption, 125, 126*f*
  stem cell research, 100, 101
Spain
  2004 top 10 GDP countries, 113*t*
  environment for research and development (R&D), 88–90
  pharmaceutical employment and R&D expenditure, 74*f*
  Profarma program, 89
  Profarma's goals supporting innovation, 90*t*
Specialty markets, chemical industry, 39–40
Start-up enterprises, pharmaceutical companies, 54
State investment, U.S. pharmaceutical industry, 65–67
State-owned companies
  employment distribution in China, 137*t*
  GDP contribution in China, 135*t*
Stem cell research, innovation for pharmaceutical R&D, 99–101
Stock price, DuPont, Monsanto, BASF, 35*f*
Strategy, short and long term, in China, 152–154
Suppliers, Chinese, 142–143
Surveys
  employment, by U.S. Bureau of Labor Analysis (BLA), 17–18
  National Science Foundation (NSF) Survey of Earned Doctorates, 62–63
  salary, by American Chemical Society, 20
Sweden
  clinical development profile for biotech companies, 87*f*
  pharmaceutical employment and R&D expenditure, 74*f*
  pharmaceutical R&D spending, 57*f*
Switzerland
  clinical development profile for biotech companies, 87*f*
  pharmaceutical employment and R&D expenditure, 74*f*
  pharmaceutical R&D spending, 57*f*

T

Taiwan, *per capita* energy consumption, 125, 126*f*

Tariffs, exports and imports, 7
Telecommunications
    China, 117–120
    fixed and mobile phone growth in China, 118f
    handset market shares, 119f
    integrated circuits, 120
    wireless market in China, 133
Televisions, China's estimated percentage of global production, 114t
Texas
    employment in U.S. pharma industry, 65f
    relationship between NIH funding and biotech companies, 66, 67t
The World is Flat, China, 120
Toffler, Alvin, *Future Shock*, 2
Total Petrochemicals
    polypropylene producer, 15
Township and village enterprise (TVE)
    China, 136–137
    employment distribution in China, 137t
Trade Related Aspects of Intellectual Property Rights (TRIPS)
    global, and innovation, 56
    public health emergencies, 102
Trusted relationships, China, 148–149

U

Union Carbide, India Bhopal accident 1984, 24
United Kingdom (UK)
    2004 top 10 GDP countries, 113t
    Association of British Pharmaceutical Industry (ABPI), 80, 82
    BioIndustry Association (BIA), 81
    Bioscience Innovation and Growth Team (BIGT), 81, 82t
    chemical industry R&D, 38
    clinical development profile for biotech companies, 87f
    environment for research and development (R&D), 80–83
    pharmaceutical employment and R&D expenditure, 74f
    Pharmaceutical Industry Competitiveness Taskforce (PICTF), 80, 83
    pharmaceutical production, 84f
    pharmaceutical R&D spending, 57f
    stem cell research, 99, 100
United States
    2004 top 10 GDP countries, 113t
    Census Bureau data, 3, 4f
    chemical industry research and development (R&D) spending totals, 36f
    chemistry industry, 11–16
    employment in biotech industry, 61f
    financial comparisons for biotech sectors in Europe and, 75t
    healthcare cost containment, 63–64
    key market for international pharmaceutical industry, 58–59
    manufacturing base, 2–6
    per capita energy consumption, 125, 126f
    pharmaceutical expenditures for R&D, 60f
    pharmaceutical R&D spending, 57f
    report from Committee on Science, Engineering and Public Policy, 64t
    stem cell research, 99–101
United States Patent and Trademark Office (USPTO), Chinese government, 97
Universities
    insourcing scientific and engineering talent, 21
    National Science Foundation (NSF) Survey of Earned Doctorates, 62–63

U.S. Bureau of Economic Analysis (BEA), GDP rate, 5
U.S. Bureau of Labor Analysis (BLA), monthly surveys, 17–18
U.S. visa requirements, non-U.S. student enrollments, 21

V

Venture capitalists, pharmaceutical companies, 54
Verbund (integration) concept, BASF, 32

W

Washing machines, China's estimated percentage of global production, 114*t*
Washington, relationship between NIH funding and biotech companies, 66, 67*t*

Western Europe, chemical industry R&D spending totals, 36*f*
World Health Organization (WHO)
    funding research and development, 102
    polluted cities in China, 132
The World is Flat, China, 120
World's workshop, China, 113–115
World Trade Organization (WTO)
    currency valuation and trade laws, 6–7
    export and import tariffs, 7

Y

Yuan
    currency revaluation, 6–8
    exchange rate, 9, 11

Highlights from ACS Books

Feedstocks for the Future: Renewables for the Production of Chemicals and Materials
Edited by Joseph J. Bozell and Martin K. Patel
392 pages, clothbound, ISBN 0-8412-3934-7

Chemical Engineering for Chemists
By Richard G. Griskey
352 pages, clothbound, ISBN 0-8412-2215-0

Advances in Microbial Food Safety
Edited by Vijay K. Juneja, John P. Cherry, and Michael H. Tunick
360 pages, clothbound, ISBN 0-8412-3915-0

A Practical Guide to Combinatorial Chemistry
By Anthony W. Czarnik and Sheila H. DeWitt
462 pages, clothbound, ISBN 0-8412-3485-X

Biogeochemistry of Chelating Agents
Edited by Bernd Nowack and Jeanne M. VanBriesen
472 pages, clothbound, ISBN 0-8412-3897-9

Medicinal Inorganic Chemistry
Edited by Jonathan L. Sessler, Susan R. Doctrow, Thomas J. McMurry, and Stephen J. Lippard
464 pages, clothbound, ISBN 0-8412-3899-5

A Lifetime of Synergy with Theory and Experiment
Andrew Streitwieser, Jr.
320 pages, clothbound, ISBN 0-8412-1836-6

For further information contact:
Order Department
Oxford University Press
2001 Evans Road
Cary, NC 27513
Phone: 1-800-445-9714 or 919-677-0977
Fax: 919-677-1303

Bestsellers from ACS Books

The ACS Style Guide: A Manual for Authors and Editors (2nd Edition)
Edited by Janet S. Dodd
470 pp; clothbound ISBN 0–8412–3461–2; paperback ISBN 0–8412–3462–0

Reagent Chemicals: Specifications and Procedures: Tenth Edition
By ACS Committee on Analytical Reagents
816 pp; clothbound ISBN 0–8412–3945–2

Advances in Arsenic Research: Integration of Experimental and Observational Studies and Implications for Mitigation
Edited by Peggy A. O'Day, Dimitrios Vlassopoulos, Xiaoguang Meng, and Liane G. Benning
446 pp; clothbound ISBN 0–8412–3913–4

Chemical Activities (student and teacher editions)
By Christie L. Borgford and Lee R. Summerlin
330 pp; spiralbound ISBN 0–8412–1417–4; teacher edition,
ISBN 0–8412–1416–6

Chemical Demonstrations: A Sourcebook for Teachers, Volumes 1 and 2,
Second Edition
Volume 1 by Lee R. Summerlin and James L. Ealy, Jr.
198 pp; spiralbound ISBN 0–8412–1481–6
Volume 2 by Lee R. Summerlin, Christie L. Borgford, and Julie B. Ealy
234 pp; spiralbound ISBN 0–8412–1535–9

The Internet: A Guide for Chemists
Edited by Steven M. Bachrach
360 pp; clothbound ISBN 0–8412–3223–7; paperback ISBN 0–8412–3224–5

Laboratory Waste Management: A Guidebook
ACS Task Force on Laboratory Waste Management
250 pp; clothbound ISBN 0–8412–2735–7; paperback ISBN 0–8412–2849–3

Metal-Containing and Metallosupramolecular Polymers and Materials
Edited by Ulrich S. Schubert, George R. Newkome, and Ian Manners
598 pp; clothbound ISBN 0–8412–3929–0

For further information contact:
Order Department
Oxford University Press
2001 Evans Road
Cary, NC 27513
Phone: 1-800-445-9714 or 919-677-0977

More Best Sellers from ACS Books

Microwave-Enhanced Chemistry: Fundamentals, Sample Preparation, and Applications
Edited by H. M. (Skip) Kingston and Stephen J. Haswell
800 pp; clothbound ISBN 0-8412-3375-6

Fire and Polymers IV: Materials and Concepts for Hazard Prevention
Edited by Charles A. Wilkie and Gordon L. Nelson
436 pp; clothbound ISBN 0-8412-3948-7

Ionic Liquids as Green Solvents: Progress and Prospects
Edited by Robin D. Rogers and Kenneth R. Seddon
614 pp; clothbound ISBN 0-8412-3856-1

Fermentation Biotechnology
Edited by Badal C. Saha
300 pp; clothbound ISBN 0-8412-3845-6

Chemometrics and Chemoinformatics
Edited by Barry K. Lavinex
216 pp; casebound ISBN 0-8412-3858-8

Polymeric Drug Delivery I: Particulate Drug Carriers
Edited by Sönke Svenson
352 pp; clothbound ISBN 0-8412-3918-5

Polymeric Drug Delivery II: Polymeric Matrices and Drug Particle Engineering
Edited by Sönke Svenson
390 pp; clothbound ISBN 0-8412-3976-2

Food Lipids: Chemistry, Flavor, and Texture
Edited by Fereidoon Shahidi and Hugo Weenen
248 pp; clothbound ISBN 978-0-8412-3896-1

Herbs: Challenges in Chemistry and Biology
Edited by Mingfu Wang, Shengmin Sang, Lucy Sun Hwang, and Chi-Tang Ho
384 pp; clothbound ISBN 978-0-8412-3930-2

For further information contact:
Order Department
Oxford University Press
2001 Evans Road
Cary, NC 27513
Phone: 1-800-445-9714 or 919-677-0977